Transition Tasks
for Mathematics
Grade 8

WALCH EDUCATION

1 2 3 4 5 6 7 8 9 10
ISBN 978-0-8251-7001-0

J. Weston Walch, Publisher
Portland, ME 04103
www.walch.com

Printed in the United States of America

Table of Contents

Introduction

Use these engaging problem-solving tasks to help transition your mathematics program to the knowledge and skills required by the Common Core State Standards for Mathematics.

This collection of tasks addresses some of the new, rigorous content found in the Common Core State Standards (CCSS) for eighth grade. The tasks support students in developing and using the Mathematical Practices that are a fundamental part of the CCSS. You can implement these tasks periodically throughout the school year to infuse any math program with the content and skills of the CCSS.

These tasks generally take 30 to 45 minutes and can be used to replace class work or guided practice during selected class periods. Depending on the background knowledge and structure of your class, however, the tasks could take less or more time. To aid with your planning, tasks are divided into two parts. This flexible structure allows you to differentiate according to your students' needs—some classes or advanced students may need only one class period for both parts, while others may need to defer Part 2 for another day or altogether. Use your own judgment regarding the amount of time your students will need to complete Parts 1 and 2. Another strategy for compressing the time necessary to complete a task is to divide the problems or calculation associated with a task among students or small groups of students. Then students can "pool" their information and proceed with solving the task.

Each Transition Task is set in a meaningful real-world context to engage student interest and reinforce the relevance of mathematics. Each is tightly aligned to a specific standard from the Grade 8 CCSS. The tasks provide Teacher Notes with Implementation Suggestions that include ideas for Introducing, Monitoring/Facilitating, and Debriefing the tasks in order to engage students in meaningful discourse. Debriefing the tasks helps students develop and enhance their understanding of important mathematics, as well as their reasoning and communication skills. The Teacher Notes also offer specific strategies for Differentiation, Technology Connections, and Recommended Resources to access online.

Student pages present the problem-solving tasks in familiar and intriguing contexts, and require collaboration, problem solving, reasoning, and communication. You may choose to assign the tasks with little scaffolding (by removing the sequence of steps/questions after the task), or with the series of "coaching" questions that currently follow each task to lead students through the important steps of the problem.

We developed these Transition Tasks at the request of math educators and with advice and feedback from math supervisors and middle-school math teachers. Please let us know how they work in your classroom. We'd love suggestions for improving the tasks, or topics and contexts for creating additional tasks. Visit us at www.walch.com, follow us on Twitter (@WalchEd), or e-mail suggestions to customerservice@walch.com.

Mowing Lawns

Common Core State Standard

Know that there are numbers that are not rational, and approximate them by rational numbers.

8.NS.1. Know that numbers that are not rational are called irrational. Understand informally that every number has a decimal expansion; for rational numbers show that the decimal expansion repeats eventually, and convert a decimal expansion which repeats eventually into a rational number.

Task Overview

Background

Prior to this task, students should have had experience with operations with rational numbers. They should also have experience with ordering rational numbers and with converting decimal expansions into rational numbers. They should know that numbers that are not rational are called irrational.

Having to calculate values using a fractional multiplier in some way is a common real-world application. Being able to convert between decimal and fractional numbers can help students order different numbers. Understanding that the decimal expansion of a rational number eventually repeats is an important building block in the area of number sense.

This task is designed to give students the opportunity to work with rational numbers and their decimal expansions in a real-world context of calculating fees.

The task also provides practice with:

- operations with rational numbers
- creating and solving proportions
- comparing rational numbers

Implementation Suggestions

Students may work individually, in pairs, or in small groups to complete one or both parts of the task. Alternatively, students may meet in groups to share their results and reflect after individually completing the task, before a class-wide discussion.

Students may use one of the online resources for instruction or review on operations with rational numbers.

Introduction

Introduce the task by asking students if they have ever completed small jobs for neighbors or relatives. Did they charge a fee for the services? Was that fee based on an hourly rate or was it a flat fee? Have they ever calculated the fee based on prior jobs? What mathematical operations need to be used in order to calculate fees?

Monitoring/Facilitating the Task

Ask questions and prompt student thinking so that they:

- Recognize the mathematical operations they are using, considering which operations are required to complete the different parts of the task.
- Recall the procedures for performing operations with rational numbers.
- Consider how their answer should relate to the question before performing the actual mathematical operation.
- Recognize that a decimal expansion that terminates actually does repeat (with zeros) and fits the definition of a rational number.
- Understand that when rounding for this exercise, they round up to the nearest hundredth. Since they're calculating fees for a business, rounding down would cost the business money.

Debriefing the Task

- Have students explain the steps they used in determining the fee for each lawn described.
- Encourage students to share their results.
- Discuss the idea of decimal expansions and look at specific examples.
- Urge students to share where they had difficulty and how they resolved their questions.
- Allow students to share their own experiences with calculating fees.
- Ask students to explore how this task can relate to their life outside of the math classroom.

Answer Key

1. One possible method is to set up a proportion to solve for the unknown amount. Sample answer: $\frac{10}{\frac{1}{2}} = \frac{x}{\frac{1}{3}}$. Students could also consider a fee of $20 for mowing one acre and then divide $20 by 1/3.

2.

Lawn size (acres)	1/2	1/3	1/4	1/5	1/6	1/7	1/8	1/9	1/10
Fee as a ratio	10/1	20/3	5/1	4/1	10/3	20/7	5/2	20/9	2/1

3.

Lawn size (acres)	1/2	1/3	1/4	1/5	1/6	1/7	1/8	1/9	1/10
Fee ($)	10.00	6.67	5.00	4.00	3.34	2.86	2.50	2.23	2.00

4. One possible method is to set up a proportion to solve for the unknown amount. Sample answer: $\frac{10}{\frac{1}{2}} = \frac{x}{0.125}$. Students could also consider a fee of $20 for mowing one acre and then multiply $20 by 0.125.

5.

Lot size (acres)	0.125	0.333	1.3	0.625	2.15
Fee as a ratio	5/2	333/50	26/1	25/2	43/1

6.

Lot size (acres)	0.125	0.333	1.3	0.625	2.15
Fee ($)	2.50	6.66	26.00	12.50	43.00

7. Use a calculator to divide each ratio to verify that the decimals are equal.

Differentiation

Some students may benefit from the use of calculators during this task. Some students may benefit from the use of one of the online resources prior to or during the task. Students who complete the task early could create a fee chart using a new base fee.

Technology Connection

Students could create a spreadsheet that calculates each fee based on the price of $10 per half acre.

Choices for Students

Following the introduction, offer students the opportunity to use their own rates for tasks such as babysitting, raking leaves, or performing other jobs. Encourage students to begin with a fraction, such as $4 for 1/2 hour of babysitting.

Meaningful Context

This task makes use of the real-world context of calculating fees. Many students will have to do this at some point in their lives. This task could be expanded by having students make a brochure with sample fees advertising a service of their choice.

Recommended Resources

- Math-Science Integration
  www.walch.com/rr/CCTTG8OpsWithRationalNumbers
  This site provides information about using operations with rational numbers and converting rational numbers to fractions.
- Proportions
  www.walch.com/rr/CCTTG8Proportions
  This site offers ten practice problems requiring the use of proportions to solve for an unknown value.
- Using Rational Numbers
  www.walch.com/rr/CCTTG8UsingRationalNumbers
  This site offers detailed instructions on how to perform operations with rational numbers, along with some practice examples.

NAME:

8.NS.1 Task • The Number System
Mowing Lawns

Part 1

To promote your lawn mowing business, you have decided to create a brochure that advertises your fees for mowing lawns of various sizes. Your basic fee is $10 for every $\frac{1}{2}$ acre mowed.

1. Describe a method to determine your fee for a lawn that is $\frac{1}{3}$ of an acre.

2. You want your brochure to include your fees for common lawn sizes. Determine the fee for each of the lawn sizes listed in the table below. Write your answer as a ratio.

Lawn size (acres)	$\frac{1}{2}$	$\frac{1}{3}$	$\frac{1}{4}$	$\frac{1}{5}$	$\frac{1}{6}$	$\frac{1}{7}$	$\frac{1}{8}$	$\frac{1}{9}$	$\frac{1}{10}$
Fee as a ratio	$\frac{10}{1}$								

3. Now that you know your fee for each lawn as a ratio, use this information to determine the dollar amount for each size of lawn mowed. Write your answer as a decimal; decimals that extend beyond the hundredths place should be rounded up.

Lawn size (acres)	$\frac{1}{2}$	$\frac{1}{3}$	$\frac{1}{4}$	$\frac{1}{5}$	$\frac{1}{6}$	$\frac{1}{7}$	$\frac{1}{8}$	$\frac{1}{9}$	$\frac{1}{10}$
Fee ($)	10.00								

continued

8.NS.1 Task • The Number System
Mowing Lawns

Part 2

A nearby subdivision has heard of your business and would like a quote for each of the houses in the neighborhood. They have presented you with a lot map that includes the size of each parcel of land written in decimal form.

4. Describe a method to determine your fee for a lot of land that is 0.125 of an acre.

5. Determine the fee for each of the following lots of land. Assume that the acreage includes only areas that can be mowed. Write your answer as a ratio.

Lot size (acres)	0.125	0.333	1.3	0.625	2.15
Fee as a ratio					

6. Use the ratios you wrote in the previous table to determine the dollar amount you would charge for each lot size.

Lot size (acres)	0.125	0.333	1.3	0.625	2.15
Fee ($)					

7. How can you verify that the rational number (in this case, the ratio) is equivalent to the decimal number?

How Many Times in a Millennium?

Common Core State Standard

Work with radicals and integer exponents.

8.EE.3. Use numbers expressed in the form of a single digit times an integer power of 10 to estimate very large or very small quantities, and to express how many times as much one is than the other. For example, estimate the population of the United States as 3 times 10^8 and the population of the world as 7 times 10^9, and determine that the world population is more than 20 times larger.

Task Overview

Background

This task is designed to illustrate the multiple ways a number can be expressed in written form. Students will collect data based on repetition of a kinesthetic act. Students then use that value to calculate, progressively, how many actions they could repeat in a millennium. Students will produce a poster that shows their action, their count for 10 seconds and their count for 1 millennium, and will express, mathematically, how many more times they can perform their action in a millennium than in 10 seconds.

The task also provides practice with:

- unit conversion
- multiplication of large numbers
- terms used for the expression of time periods
- powers of ten
- noticing patterns
- making predictions

Implementation Suggestions

- Students may work individually, in pairs, or in small groups to complete one or both parts of the task.
- Students will need materials for poster production. Posters may be produced physically or digitally.
- Allow 5–8 minutes for the introduction. Allow 10 minutes for data collection and number calculation and 10 minutes for poster production. The poster walk and final debriefing can be accomplished in 15 minutes.

Introduction

Introduce the task. Ask the students to suggest words used to express periods of time. Students should respond with common time units (second, minute, hour, day, week, month, year, decade, century). Remind students that these words indicate units of time and that time periods can be converted between units. Students should know that multiplication is used to convert from a second to a minute, a minute to a year, etc. List the following words and conversion factors on the board.

second
minute = 60 seconds
hour = 60 minutes
day = 24 hours
week = 7 days
year = 52 weeks
decade = 10 years
century = 100 years
millennium = 1,000 years

Have students observe that between decade and era the years expressed by the units given are increasing by powers of ten. Students should be familiar with the use of exponents to express powers of ten. You may want to compile a list on the board, making connections between the language common to both (i.e., 100 years is a century, 1,000 years is a millennium).

$10^0 = 1$ (year)
$10^1 = 10$ (decade)
$10^2 = 100$ (century)
$10^3 = 1{,}000$ (millennium)

Each student will have to select a kinesthetic activity (blinking, snapping, clapping, jumping, stamping their foot, etc.) that they can do multiple times in a 10-second period. Alternatively, you may suggest a specific activity or present a list for students to select from.

Allow students a trial run to practice data collection. Keep time for the class, giving a verbal signal to begin data collection at 0, then giving another verbal cue for them to stop after 10 seconds. Ask for one or two students to volunteer to demonstrate their activity and talk through the process of completing the statement: "I measured that I can <u>(insert activity)</u> <u>(insert count)</u> times in 10 seconds." (Example: I measured that I can <u>blink</u> <u>20</u> times in 10 seconds.)

Have students estimate how many times they think they could repeat the action in a millennium (1,000 years) if it were somehow possible to keep going that long. For a more realistic scenario, describe how pulsar stars emit regular radiation bursts. If a particular pulsar emits 3 bursts of radiation every 10 seconds, how many bursts would it emit in a millennium?

You may wish to model the data collection and provide an example of value conversions on the board before students begin.

Monitoring/Facilitating the Task

Ask questions and prompt student thinking so that they:

- Focus on the multiple ways of writing each value. A sample response table is included to illustrate the way students should be completing their work. Ask what has changed the number from row to row in the table. Students should respond that the multiplier used for unit conversion has made the number that many times larger. For example, the number of repetitions per hour is 60 times larger than the number of repetitions per minute.
- Realize the connection between the powers of ten and the exponents used. Ask what is happening to their value as the exponent increases. Ask them to justify their response. Students should observe that as their number grows larger, the exponents increase.
- Observe that only the exponent is changing in the power of ten representations as they calculate their repetitions for decade through millennium. Ask them to investigate the pattern they are seeing. Students should observe that each increase in the exponent is a 10-time increase in repetitions.
- Compare decimal values and powers of ten.

As students fill in responses to Part 2, ensure that they properly use the terms *measure, estimate,* and *calculate* as they discuss their work.

Allow students to complete their posters. Check their progress and remind them to include all the required aspects on their poster. Display posters in the room for a poster walk during debriefing.

Debriefing the Task

- Allow students time for a poster walk to observe the work done by their classmates. As they review the posters, ask students to think about putting them in order. Ask them to decide which activity could be done the most and the fewest times in a millennium (1,000 years).
- During the poster walk, have students write their number per millennium on the board in whole-number form.
- Explore the ordering of larger numbers. Ask students to use the board to indicate the largest and smallest repetition values expressed. Ask the students to find the largest and smallest values from the board. Students should express difficulty in making the comparison because of the number of integers written on the board.

- Have the students move their focus to the posters. Ask them to order the posters by reading the exponential notation representation. Students should observe their increased ability to order values when powers of ten are used. Have them justify their responses using the powers of ten and exponent values. Students should connect that the number with the largest power of ten is the largest number. Ask for student input for ordering the posters in the classroom, from smallest power of ten to largest.
- In summary, make sure that students understand the purpose and value of scientific notation/exponents in recording and working with very large numbers.

Answer Key

Answers will vary; all answers are sample answers.

1. I measured that I can snap my fingers three times in 10 seconds.
2. I estimate that I can snap my fingers three billion times in one millennium.
3. Sample table:

Multiplication	Integer value	Time span for repetitions	Decimal	Power of ten
3 • 6	18	1 minute	1.8 • 10	$1.8 \cdot 10^1$
18 • 60	1,080	1 hour	1.08 • 1,000	$1.080 \cdot 10^3$
1,080 • 24	25,920	1 day	2.592 • 10,000	$2.5920 \cdot 10^4$
25,920 • 7	181,440	1 week	1.8144 • 100,000	$1.81440 \cdot 10^5$
181,440 • 52	9,434,880	1 year	9.43488 • 1,000,000	$9.434880 \cdot 10^6$
9,434,880 • 10	94,348,800	1 decade	9.43488 • 10,000,000	$9.434880 \cdot 10^7$
94,348,800 • 10	943,488,000	1 century	9.43488 • 100,000,000	$9.434880 \cdot 10^8$
943,488,000 • 10	9,434,880,000	1 millennium (1000 years)	9.43488 • 1,000,000,000	$9.434880 \cdot 10^9$

4. I measured that I could snap my fingers three times in 10 seconds.
5. I estimated that I could snap my fingers three billion times in one millennium.
6. I calculated that I could snap my fingers $9.43488 \cdot 10^9$ times in one millennium.
7. I can snap my fingers 9.4 billion times more in a millennium than I can in 10 seconds.
8. Check student posters for completion and accuracy. Make sure student posters include the number of times they could complete their selected action in a millennium using the tens place representation of the value.

Differentiation

- Some students may benefit through the use of calculators during this task.
- Interested students could collect data from multiple students prior to the activity.
- Some students may wish to continue calculations beyond a millennium using higher powers of ten.
- Some students may benefit by further exploring the concept of ordering. Estimation values could be used in an expansion of this task.

Technology Connection

- Spreadsheet software could be used for generating multiples and recording data.
- Graphic design software could be used for poster production.

Choices for Students

- Interested students may choose to express large values (or smaller values) from a different source (computing terms, scientific values), or order the estimations students give.
- In place of the poster, students could produce a visual representation of the large value. For examples of this, see the MegaPenny Web site (URL listed in Recommended Resources).

Meaningful Context

Students are routinely exposed to numbers in the millions, billions, and trillions in their world. It can be difficult to visualize these numbers and put them in context with smaller values. This task offers multiple representations for large numbers.

Recommended Resources

- The MegaPenny Project
  www.walch.com/rr/CCTTG8MegaPenny
  The MegaPenny Project represents large numbers using pictures of pennies, with excellent visual representations.

- Wolfram MathWorld—Large Number
  www.walch.com/rr/CCTTG8LargeNumbers
  This site provides information on large numbers and words for various powers of ten.

NAME: ______

8.EE.3 Task • Expressions and Equations
How Many Times in a Millennium?

Part 1

Use the following information to complete the task.

second	year = 52 weeks
minute = 60 seconds	decade = 10 years
hour = 60 minutes	century = 100 years
day = 24 hours	millennium = 1,000 years
week = 7 days	

$10^0 = 1$ (year)
$10^1 = 10$ (decade)
$10^2 = 100$ (century)
$10^3 = 1{,}000$ (millennium)

Choose an action to perform, such as snapping your fingers. Count the number of times that you can perform your action in 10 seconds. Use that piece of data to complete the records below, using the same action for each.

1. I measured that I can ______
   ______ times in 10 seconds.

2. I estimate that I can ______
   ______ times in one millennium.

continued

8.EE.3 Task • Expressions and Equations
How Many Times in a Millennium?

3. Convert between different forms to fill out the information in the table. The first row is an example, using 3 times per second as a sample answer. Look at it to see how each column is to be filled in, using your own data.

Multiplication	Integer value	Time span for repetitions	Decimal	Power of ten
$3 \cdot 6$	18	1 minute	$1.8 \cdot 10$	$1.8 \cdot 10^1$
		1 minute		
		1 hour		
		1 day		
		1 week		
		1 year		
		1 decade		
		1 century		
		1 millennium (1,000 years)		

8.EE.3 Task • Expressions and Equations
How Many Times in a Millennium?

Part 2

Complete the following sentences using the power of ten representations from the table.

4. I measured that I could ______________________________

   ______________ times in 10 seconds.

5. I estimated that I could ______________________________

   ______________ times in one millennium.

6. I calculated that I could ______________________________

   ______________ times in one millennium.

7. I can ______________________________

   ______________ times more in a millennium than I can in 10 seconds.

8. Design and create a poster for display in your classroom. Make sure your poster includes the number of times you can complete your action in a millennium using scientific notation.

How Much Money Will I Save?

Common Core State Standard

Understand the connections between proportional relationships, lines, and linear equations.

8.EE.5.1. Graph proportional relationships, interpreting the unit rate as the slope of the graph. Compare two different proportional relationships represented in different ways. ...

Task Overview

Background

This task has students graph proportional relationships and interpret the unit rate as the slope of the graph. Each group will calculate a given discount amount for several items of furniture. They will then present their cost and discount information in a graph, observing a linear relationship. Students will calculate the slope of the relationship and make connections between the slope, the unit rate, and the percentage discount.

Students should have prior experience with calculating percent off when given a value and percent. They should also have experience with finding slope either by using a graph and the rise over run method, or calculating slope using a formula.

Implementation Suggestions

Students should be arranged into five groups for this task—one for each possible discount card. No calculators should be used at the end of this task, although they may be helpful when completing problem 1.

Introduction

Ask students if they have ever gone shopping during a sale. Ask them about their strategy for knowing if they can afford what they want to buy. What kind of discounts did they commonly see? How could they calculate percentage off without a calculator?

If necessary, remind students that percentage represents a proportional relationship. No matter what quantity you have been given, discounting a percent off will result in a number that is proportional to the original price. Ask students where they have seen proportional relationships before. Remind students how to set up a proportion to solve for the discount amount.

If appropriate, explain that in this problem they are relating the discount to the original price of the furniture. This is a proportional relationship; it passes through the origin, and if the cost of the furniture is doubled or tripled, the discount received is also doubled or tripled.

Organize students into five groups and assign each group one discount amount for problem 1: 5%, 10%, 15%, 25%, or 50%.

Monitoring/Facilitating the Task

Ask questions and prompt student thinking so that they:

- Realize that the percent off and the values they calculate in problems 5 and 7 represent the **proportion constant**.
- Recognize that they have represented the same relationship in three ways—in a table, in a graph, and in an equation.
- Observe that the information they amassed in problem 8 can be applied to other values.
- See the pattern, "If I save $2.50 off every $10.00 I spend, then I must get $5.00 off for every $20 I spend."
- Understand that this is the **percent proportional relationship**.

Debriefing the Task

- Have each group present their results—the percent discount, a table showing the cost and the discount, a graphic representation, and an equation—to the class.
- Lead a discussion to compare and contrast each table, graph, and equation.
- Students should observe that in each situation the graph, table, and equation are similar and only differ in the discount amount.
- Display the graph from the answer key for problem 2, with all five discounts shown.
- Have the students compare the similarities and differences in each representation.
- Ask students where additional relationships would fit on the graph. Word these relationships in several ways. Where would a discount of 17% fit in? What if the unit rate was 0.75? Where would a line with a slope of 0.6 appear?
- Draw connections among the multiple representations used throughout the task.

Answer Key

1. Answers will vary based on discounts assigned. Table of all possible results:

Item	Original cost	5% discount	10% discount	15% discount	25% discount	50% discount
Chair	$345.00	$17.25	$34.50	$51.75	$86.25	$172.50
Coffee table	$113.00	$5.65	$11.30	$16.95	$28.25	$56.50
Couch	$848.00	$42.40	$84.80	$127.20	$212.00	$424.00
End table	$58.00	$2.90	$5.80	$8.70	$14.50	$29.00
Rug	$92.00	$4.60	$9.20	$13.80	$23.00	$46.00
Standing lamp	$77.00	$3.85	$7.70	$11.55	$19.25	$38.50
Table lamp	$46.00	$2.30	$4.60	$6.90	$11.50	$23.00
TV stand	$298.00	$14.90	$29.80	$44.70	$74.50	$149.00

2. $y = x \bullet \%$, where y is the discount, x is the cost, and % is the percent off

3. Graphs will vary based on discounts assigned. Graph of all possible plotted points:

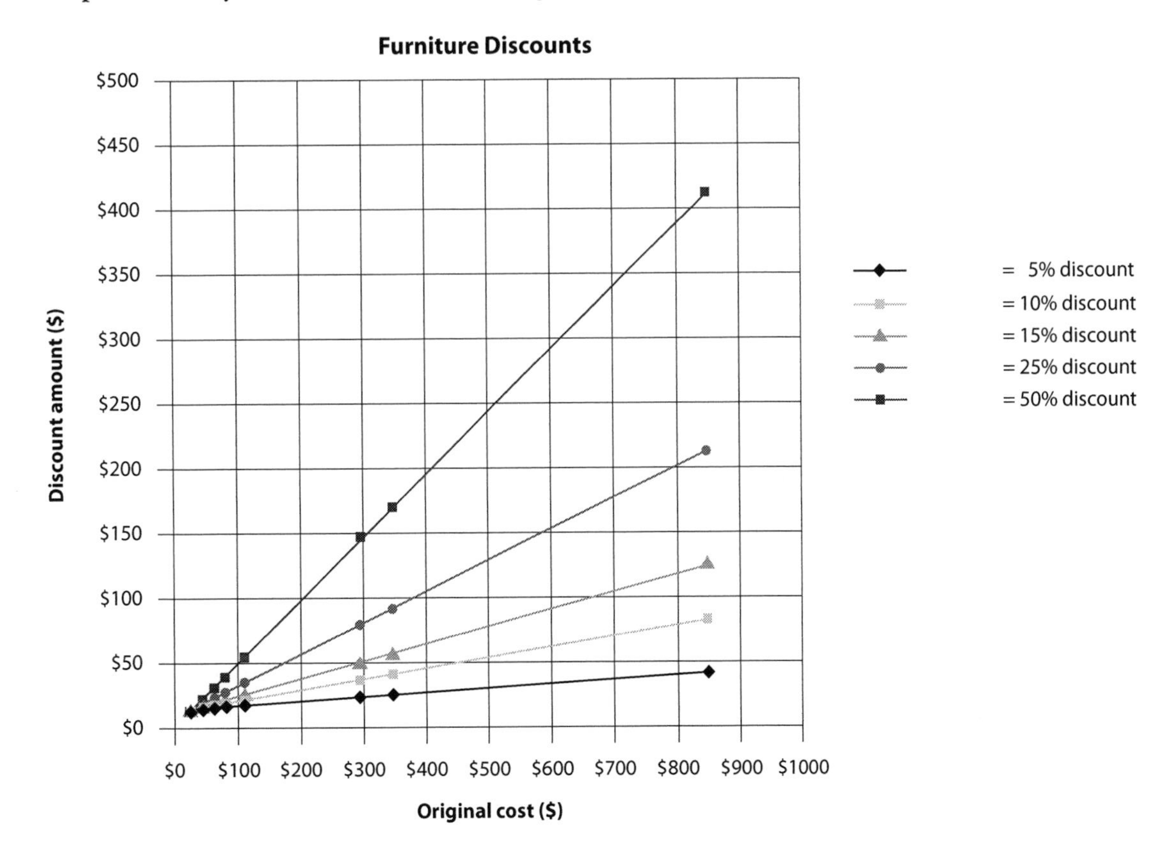

4. The graph shows a linear or proportional relationship. This is evident because the line is straight.
5. Slopes will vary based on the discount assigned. Possible slopes: 0.05, 0.10, 0.15, 0.25, 0.50
6. The slope and the percent discount are the same.
7. Unit rates will vary based on the discount assigned. Possible rates: 0.05, 0.10, 0.15, 0.25, 0.50
8. The slope, the percent discount, and the unit rate are the same.
9. Answers will vary based on discounts assigned. Use the following table of all possible results to check the completed sentences.

	Money spent			
Percentage discount	$1.00	$10.00	$100.00	$1,000.00
5%	$0.05	$0.50	$5.00	$50.00
10%	$0.10	$1.00	$10.00	$100.00
15%	$0.15	$1.50	$15.00	$150.00
25%	$0.25	$2.50	$25.00	$250.00
50%	$0.50	$5.00	$50.00	$500.00

10. Answers will vary, but estimates should be between $90 and $100. Check explanations for reasonableness.
11. Answers will vary, but should be around $188. Check explanations for reasonableness.
12. Answers will vary, but estimates should be between $275 and $300. Check explanations for reasonableness.
13. Answers will vary, but estimates should be between $450 and $500. Check explanations for reasonableness.
14. Answers will vary, but estimates should be around $940. Check explanations for reasonableness.
15. 15%
16. 50%
17. 5%

Differentiation

Some students may benefit from the use of smaller values in the initial task.

Technology Connection

Students could produce graphs and tables in a spreadsheet application, such as Microsoft Excel or Calc from OpenOffice.org.

Choices for Students

Students could replicate the shopping problem using different scenarios to challenge their classmates to determine how much was saved.

Meaningful Context

Many students are challenged by calculating discounts and tax when in a shopping scenario.

Ask students to use the proportional nature of percent to track tax amount next time they are shopping.

Have students research how sales tax works, what the local and state sales tax rates are, and how sales tax values remain proportional whether the purchase is for $1.00 or $1,000.

Recommended Resources

- Bagel Algebra
  www.walch.com/rr/CCTTG8BagelAlgebra
  This activity may be used to review or illustrate proportional reasoning.
- Percent and Proportions
  www.walch.com/rr/CCTTG8PercentProportions
  This site offers a lesson on percent proportions, example problems, and Web-based practice problems.
- Sales Tax by State
  www.walch.com/rr/CCTTG8StateSalesTaxes
  This resource from Living American.com is geared toward new immigrants. It offers a clear explanation of sales taxes, along with a chart of sales tax rates for each state.

8.EE.5.1 Task • Expressions and Equations
How Much Money Will I Save?

Introduction

A local furniture store is giving away a scratch-off card to each customer for discounts on purchases. When you scratch the card, you reveal the percent off you will receive on your next purchase. The store offers discounts of 5%, 10%, 15%, 25%, and 50%.

Your family is shopping for new living room furniture. You decide to purchase the items listed in the following table.

Item	Original cost
Chair	$345.00
Coffee table	$113.00
Couch	$848.00
End table	$58.00
Rug	$92.00
Standing lamp	$77.00
Table lamp	$46.00
TV stand	$298.00

Part 1

1. You scratch your card and get a discount of _______% (fill in the percent of discount assigned by your teacher). Calculate the amount you will save on each purchase using that discount percentage. Record the amount of discount for each purchase in the table below.

Item	Original cost	Discount amount
Chair	$345.00	
Coffee table	$113.00	
Couch	$848.00	
End table	$58.00	
Rug	$92.00	
Standing lamp	$77.00	
Table lamp	$46.00	
TV stand	$298.00	

8.EE.5.1 Task • Expressions and Equations
How Much Money Will I Save?

2. Write an equation to describe the data in the table. Use x to represent the cost and y to represent the discount.

3. Plot the points from your table on the graph below. Label the axes of the graph.

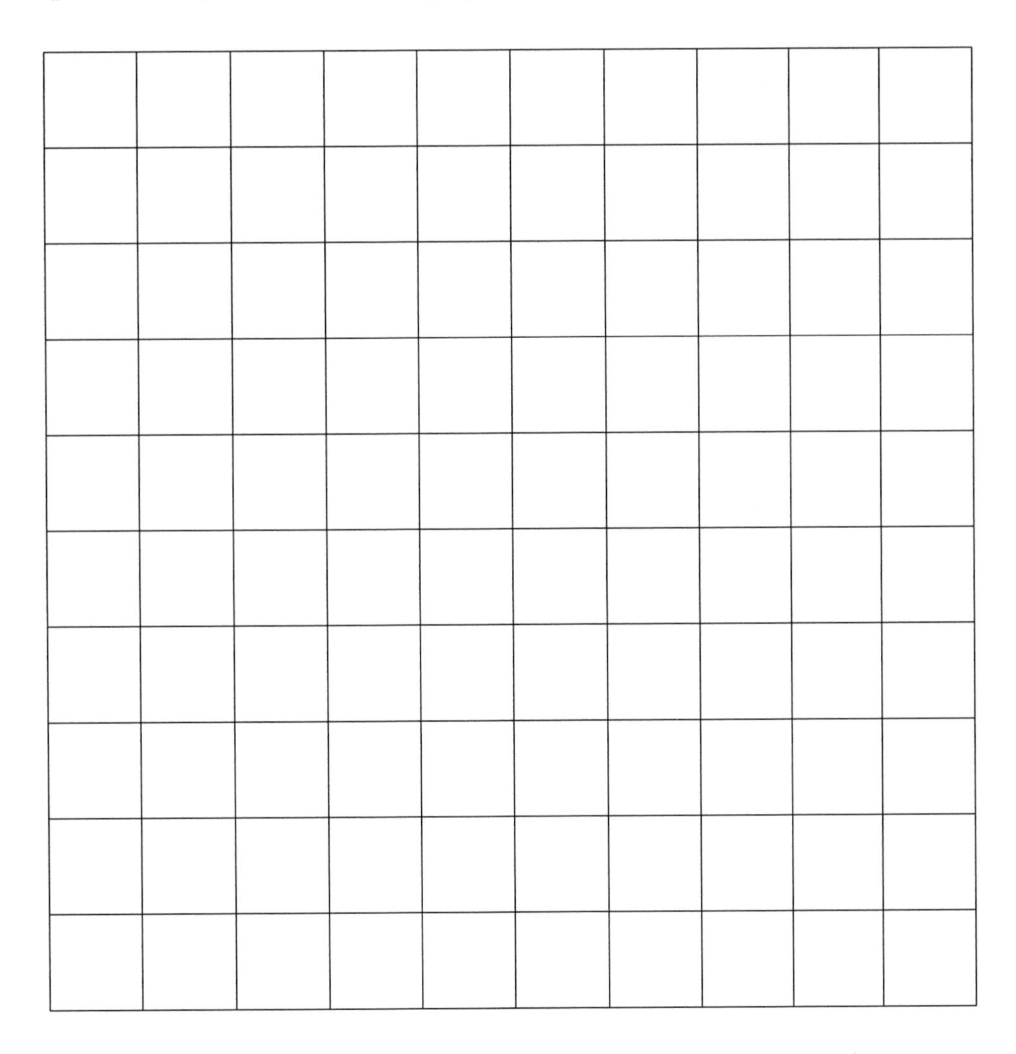

4. What type of relationship is shown in your graph? How can you tell?

continued

5. Calculate the slope of the line shown on your graph.

6. How does the slope of your line relate to the percent discount you were assigned?

7. Calculate the **unit rate of the proportion**. This is the amount of the discount if you were to purchase an item costing $1.00.

8. How does the unit rate of the proportion relate to the slope of the line and the percent discount you were assigned?

continued

9. For each sentence below, fill in the discount percentage you were assigned. Then use the unit rate that you calculated in problem 7 to determine how much you would save for each amount spent.

   a. With a discount of ______%, we will save ____________ for each $1.00 we spend.

   b. With a discount of ______%, we will save ____________ for each $10.00 we spend.

   c. With a discount of ______%, we will save ____________ for each $100.00 we spend.

   d. With a discount of ______%, we will save ____________ for each $1,000.00 we spend.

Your purchases total $1,877.00 before discounts. Use the unit rate to estimate your savings for each of the discount coupons. DO NOT USE A CALCULATOR. For each discount, explain how you estimated the amount.

10. Discount amount: 5%

11. Discount amount: 10%

continued

12. Discount amount: 15%

13. Discount amount: 25%

14. Discount amount: 50%

Part 2

Your dad went to the same store three different times to purchase additional items for the living room, dining room, and kitchen. He loves math and saving money, so he created a challenge for you to determine the different discounts he received on each purchase.

15. Use the graph your dad produced to determine the discount amount of the scratch-off card he used to make some small living room purchases.

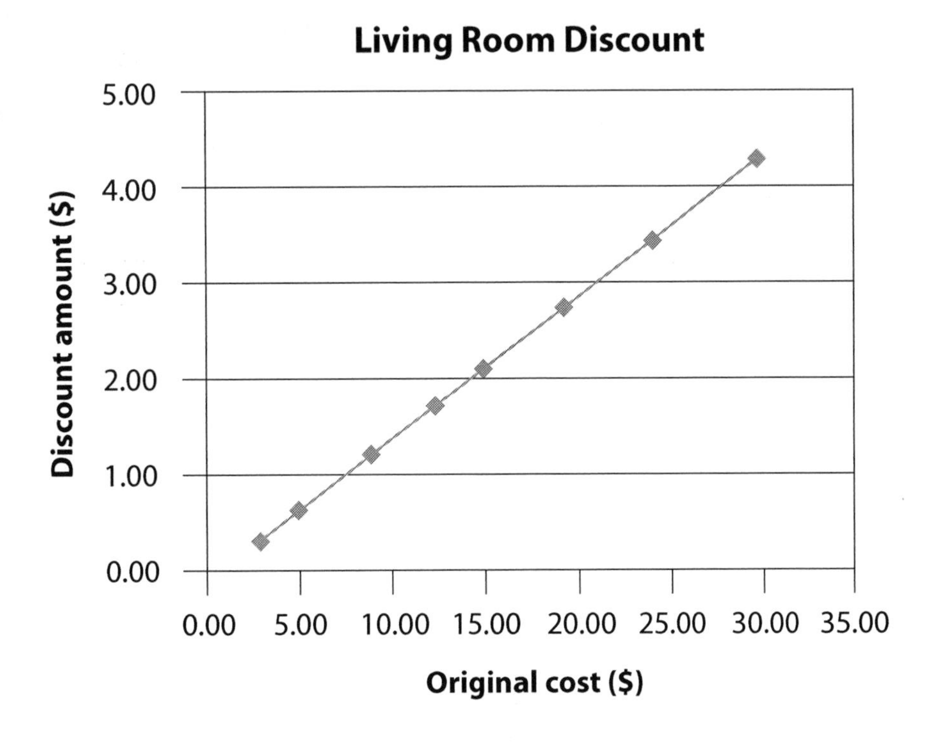

Discount amount: _______%

continued

8.EE.5.1 Task • Expressions and Equations
How Much Money Will I Save?

16. Use the table your dad produced to determine the discount amount of the scratch-off card he used to make dining room purchases.

Dining room discount	$217.50	$8.00	$48.00	$3.00	$29.50	$36.50	$4.00	$7.50
Original cost	$435.00	$16.00	$96.00	$6.00	$59.00	$73.00	$8.00	$15.00

Discount amount: _______%

17. Use the equation your dad produced to determine the discount amount of the scratch-off card he used to make kitchen purchases.

Kitchen discount = 0.05 • original price

Discount amount: _______%

Rolling Tennis Balls

Common Core State Standard

Understand the connections between proportional relationships, lines, and linear equations.

8.EE.5.2. Compare two different proportional relationships represented in different ways. For example, compare a distance-time graph to a distance-time equation to determine which of two moving objects has greater speed.

Task Overview

Background

Proportional relationships are common real-world applications. Understanding the different ways these relationships can be represented is an important concept. Being able to compare graphs and their corresponding equations is a crucial learning goal in algebra.

This task is designed to give students the opportunity to explore the connection between proportional relationships, lines, and linear equations using time and distance data they have collected themselves. They will record how long it takes a tennis ball to travel a certain distance and use the data to create graphs and equations.

Alternatively, if you are concerned about time or the availability of materials, you can provide students with completed data tables and have them skip to problem 3 on the student worksheet.

The task also provides practice with:

- measuring
- rounding
- recording data
- creating graphs
- writing equations

Implementation Suggestions

Students ideally should work in groups of three or four. Each student will assume the role of timekeeper, ball roller, or data recorder for each trial. Students will need to have measuring tapes or yardsticks to measure the distances, a stopwatch to track the time, and tennis balls to roll.

Introduction

The task should be introduced following instruction on proportional relationships, graphing lines, and writing equations from a data set. Students should be able to identify independent and dependent variables.

If necessary, begin the task by having students recall the relationship between time, distance, and rate of speed ($d = r \times t$), and how to determine rate of speed ($r = d/t$). Review the process of creating a graph from a data set. Make sure students understand the importance of completely labeling their graphs (so others can understand what the graph represents). Discuss which point all of the lines will go through (the origin) and why (because the starting time and distance for all trials is 0 seconds and 0 feet).

Make sure students know how to use the measuring tape or yardstick and the stopwatch to accurately take the measures. Include a reminder of rounding rules. Discuss the importance of identifying the units of measure (feet or meters and seconds).

Prepare students for work in groups of three or four. Each student should perform each task at least once. Each student in the group should record the data on their own sheet, create their own graph, calculate their own rates of speed, and write their own equations.

Separate students into groups. Provide each group with the materials they will need to make the measurements, record the data, and create their graphs.

Monitoring/Facilitating the Task

Ask questions and prompt student thinking so that they:

- Make accurate measurements.
- Assume each role at least once during the task.
- Record their group's data correctly.
- Graph all four trials on one set of axes.
- Use appropriate units of measure in their data tables and graphs.
- Carefully compare their graphs and their equations.
- Think about other examples of proportional relationships (for example, height and age, hours worked and amount earned, and automobile value over time).

Debriefing the Task

- Upon completion of the task, students should share their results.
- Ask students to describe the process they used in creating their graphs and writing their equations.
- Ask students to discuss which variables they chose and why they chose them. Discuss why it is convenient to use conventional variables, such as r for rate of speed, t for time, and d for distance (since others are familiar with these designations and they are "memorable").
- Ask students to describe the connection between their graphs and their equations. What are the patterns? How are they connected? What role does slope have in this task? (The slope of the line represents the rate of speed; the steeper the slope, the greater the rate of speed. In the equation, the rate of speed is the coefficient of the variable representing time.)
- Discuss the fact that the four trials are not necessarily related to one another. In each one, *distance* = *rate of speed* × *time*, but there is no standardization for rate of speed and it will likely vary across the trials.
- Ask students to share any difficulties they had with this task. Encourage them to describe how they addressed these difficulties.
- Encourage students to share their examples of other proportional relationships. Why did they choose these examples?

Answer Key

Actual answers will vary depending upon the data collected. Students fill out the table below in response to problems 1, 2, 3, and 6. Sample answers are given here as a guide. Sample completed table:

	Distance	Time	Rate of speed	Equation
Trial 1	20 ft	4.2 sec	4.76 ft/sec	$d = 4.76t$
Trial 2	30 ft	3.7 sec	8.11 ft/sec	$d = 8.11t$
Trial 3	35 ft	4.6 sec	7.61 ft/sec	$d = 7.61t$
Trial 4	40 ft	3.9 sec	10.26 ft/sec	$d = 10.26t$

1. Check data table for recorded distances.
2. Check data table for recorded times.
3. Rate of speed = distance/time; division; check student calculations and refer to the data table for rate results.

4. Time is the independent variable and distance is the dependent variable. Sample graph:

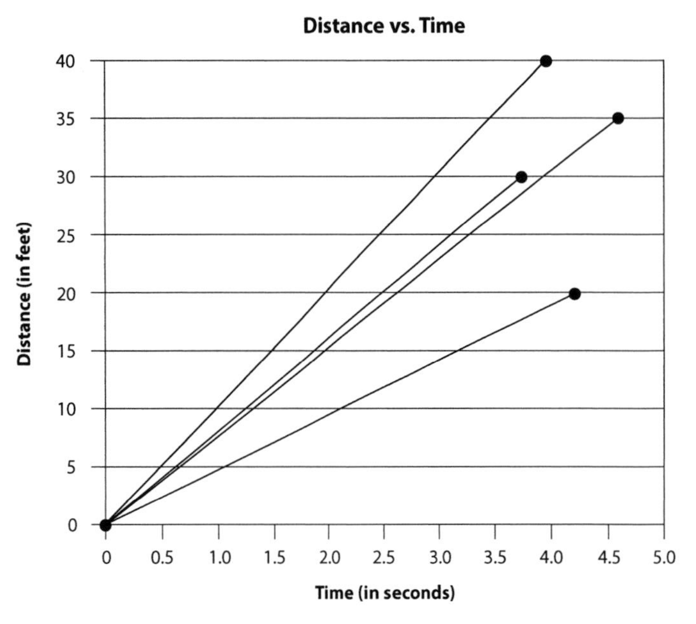

5. Each slope represents the ball's rate of speed for each trial. The line with the steepest slope corresponds to the fastest tennis ball. The line with the most gradual slope represents the slowest tennis ball.
6. Check data table for equations.
7. Students may indicate they chose the variables d and t because they are recognizable and easy to remember; d stands for distance in feet and t stands for time in seconds.
8. Yes—the equation with the largest coefficient corresponds to the tennis ball with the greatest rate of speed; the equation with the smallest coefficient represents the tennis ball with the slowest rate of speed.
9. The line with the steepest slope corresponds to the equation with the largest coefficient; the line with the most gradual slope corresponds to the equation with the smallest coefficient.
10. Sample answer: the relationship between time and the amount of water lost from a leaking faucet

Differentiation

Some students may benefit from the use of calculators during this task. Some may find it helpful to complete one of the online instruction/review activities prior to completing this task. Some students may be more successful if they are given data, rather than asked to collect it themselves.

Students who complete the task early could perform additional trials and/or introduce new variables (different ball sizes, wrapping the tennis ball in paper and tape, etc.), creating additional equations and graphs to illustrate.

Technology Connection

Students could use a spreadsheet program (Excel or Google Spreadsheet) to create a data table and graph and compare it to their hand-drawn graph.

Choices for Students

Following the introduction, offer students the opportunity to collect slightly different data. They could time balls dropped from different heights or roll different types of balls.

Meaningful Context

This task makes use of actual data that students collect themselves. The task allows students to make accurate measurements, calculate rates of speed, and compare graphs and equations. The relationship of *distance* = *rate of speed* × *time* is found in many real-world applications; students should become very familiar with it.

Recommended Resources

- Interactive Position vs. Time Graph
  www.walch.com/rr/CCTTG8DistanceTime
  Students guide a spaceship to its mother ship in order to explore the relationship between distance and time in this interactive program.
- Learning Math: Proportional Reasoning
  www.walch.com/rr/CCTTG8ProportionalReasoning
  This site provides video- and Web-based activities to review basic concepts of proportional reasoning.
- NASA Lesson: Getting Off the Ground into the "Smart Skies"!
  www.walch.com/rr/CCTTG8SmartSkies
  This alternative lesson/activity explores the connection between proportional relationships and lines.
- Plain Graph Paper PDF Generator
  www.walch.com/rr/CCTTG8GraphPaper
  Customize this downloadable graph paper to various paper dimensions, line weights, grid sizes, and colors.

8.EE.5.2 Task • Expressions and Equations
Rolling Tennis Balls

Part 1

You are going to calculate the rate of speed of rolling tennis balls by measuring the time it takes the ball to roll a certain distance. You will record the data, create a chart, and create equations to represent the rate of speed of the ball. Then you will compare the graphs and the equations to determine the relationship between them.

1. In your group, mark 4 different distances in the hallway or in your classroom. Designate who will be the ball roller, the timekeeper, and the data recorder for each trial, taking turns with each job. Record your distances in the table below.

	Distance	Time	Rate of speed	Equation
Trial 1				
Trial 2				
Trial 3				
Trial 4				

2. Take turns rolling the tennis ball. For each trial, try rolling the ball more slowly or more quickly. Record the time it takes to roll the distances you have marked. Round each time to the nearest tenth of a second. Each person in your group should copy the data to his or her own data sheet. Be sure to include the appropriate units of measure.

3. How could you determine the ball's rate of speed for each trial? Which operation is used? Fill out the data table with your calculations. Be sure to use the correct units of measure.

8.EE.5.2 Task • Expressions and Equations
Rolling Tennis Balls

4. Think about the data you gathered. Create a graph showing the relationship between time and distance. Which measure would represent the independent variable? Which would represent the dependent variable? Plot all 4 sets of data on the same set of axes. Label your graph.

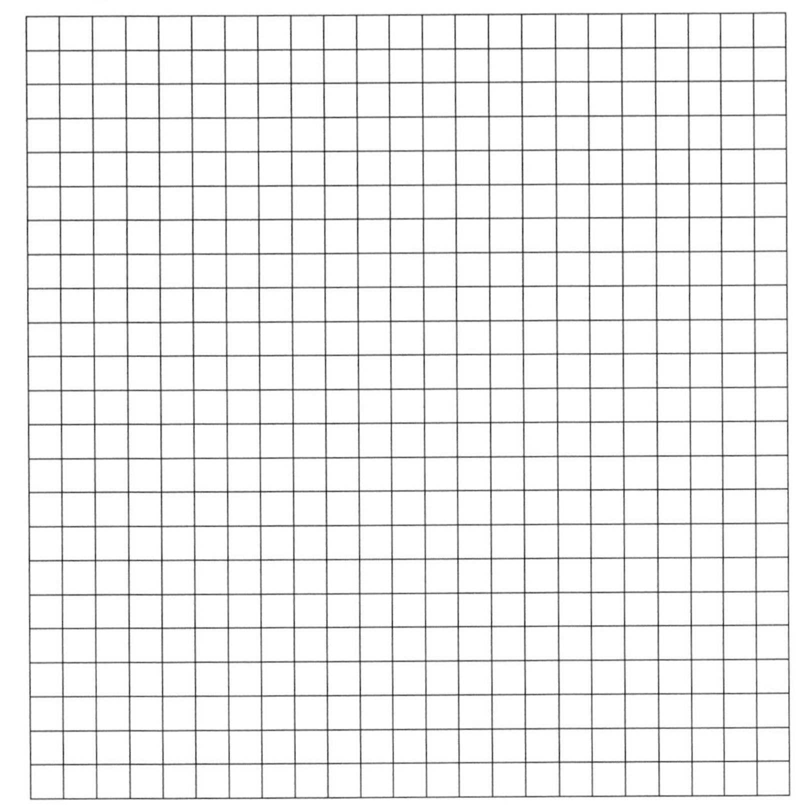

5. Consider your graph. What do the slopes of the lines represent? Can you tell which tennis ball was moving the fastest or slowest by looking at the graph? How?

continued

Part 2

6. Look at your data table again. In the last column of the data table in Part 1, write an equation for each trial that would represent the relationship between distance and time.

7. Think about the equations you have written. How did you determine which variables to use? What do the variables represent?

8. Consider the equations you have written. Can you tell which tennis ball was moving the fastest or slowest by looking at the equations? How?

9. Look at the equations and your graph. Can you see a connection between the equations you have written and the lines on the graph? Describe fully.

10. Can you think of another relationship between two things that could be represented by a similar equation or a graph? Describe fully.

8.EE.7(a) Task • Expressions and Equations

Cell Phone Plans

Instruction

Common Core State Standard

Analyze and solve linear equations and pairs of simultaneous linear equations.

8.EE.7a. Give examples of linear equations in one variable with one solution, infinitely many solutions, or no solutions. Show which of these possibilities is the case by successively transforming the given equation into simpler forms, until an equivalent equation of the form $x = a$, $a = a$, or $a = b$ results (where a and b are different numbers).

Task Overview

Background

Students often struggle when working with equations that contain infinitely many or no solutions, despite being able to solve a linear equation with one solution. This task helps students internalize the structure and the meaning of equations with infinitely many or no solutions.

The task also provides practice with:

- transforming equations
- solving equations with one solution

Implementation Suggestions

Students may work individually or in pairs for Part 1 and in pairs or small groups to complete Part 2. Part 1 should take only 5 minutes and Part 2 should take about 25 minutes to complete.

Students can present their plans at the end of Part 2 to the whole class. Alternatively, students may meet in different groups to share their results and reflect after completing the task, before a class-wide discussion.

Introduction

Introduce the task by asking students about how to solve linear equations. Ask students if they have ever encountered an equation that had infinitely many solutions or no solutions.

Question students about their knowledge of how cell phone plans and charges for airtime work. Providers usually charge a flat monthly fee and assess a fee per minute if users go over their allotted time. Ask students if they have ever been involved with comparing and choosing plans for cell phone service providers. Ask students how they might use math to compare plans. Careful analysis will reveal the differences and similarities of the plans.

Monitoring/Facilitating the Task

Ask questions and prompt student thinking so that they:

Part 1

- Realize the connection between the verbal statement and the algebraic expressions.
- Recognize the mathematical operations they are using. Make sure that students articulate where and when they are using inverse operations to isolate the variable in transforming equations.
- Understand why plans are set equal to each other for comparing.
- Report and consider what their typical cell phone usage is in minutes to help in determining which plan is better.
- Defend their responses. Make sure that students articulate how they used their calculations to answer which cell phone plan they chose.

Part 2

- Recognize that the rate in minutes is the coefficient of the variable t, and the monthly fee is the constant.
- Realize that if a discount is $x\%$, then what is paid is $100\% - x\%$.
- Apply skills for working with decimals and percents, particularly with converting between the two and performing operations with them.
- Understand how to use the distributive property.
- Think about the structure of the equations and any patterns that emerge, such as instances in which a variable is present on both sides of the equation—sometimes the variable adds out and the result is an equation that is true (infinite solutions) and sometimes the result is an equation that is false (no solution).
- Explain their reasoning for creating the cell phone plans, such as choosing rates that are the same or different, and using constants that are more or less than the initial plan.

Debriefing the Task

Part 1

- Students will be solving linear equations with a variable on each side. They will set two equations equal to each other and solve for the time, t. The solutions represent the number of minutes used that indicate when the plans are equal in cost.
- Students must recognize what the solution means in terms of the problem's context. They may have difficulty translating the numerical solution into contextual meaning. The time, t, in minutes represents the number of minutes used when the plans are equal in cost; for fewer minutes used, Plan A is more economical, and after 200 minutes the current plan is better.
- Question 2 solutions may vary; either plan would be economical depending on if students use the phone less than 200 minutes (then Plan A is the better choice) or more than 200 minutes (so the current plan is more economical). Elicit responses so both answers are represented.

Part 2

- For question 3, have students discuss how and why they chose the numbers they did for their plans. Each group's numbers may differ, but the equations should be similar, with a decimal for the coefficient and a number from 20 to 200 as the constant. Ask for different examples.
- For questions 4–6, ask students to select a person from their group to present their created plans to the whole class. Encourage other students to ask questions to the presenting group about the strategies they used and have the students comment on any differences in strategies.
- For question 7, encourage discussion of how students determined the algebraic criteria for creating the plans; for example, to create two plans that have no solution means that the variable adds out and the remaining statement is false.
- Encourage students to discuss what one solution, no solutions, and infinite solutions each mean in this context. One solution means that one plan has a clear economical advantage over the other after a certain amount of minutes are used. The solution will represent that number of minutes. No solutions means that one plan is better than the other for any amount of minutes used, and infinite solutions means that the plans are identical.
- For question 8, students are asked to make a generalization and synthesize their observations. Students might need help seeing the pattern of the structure of equations with one solution, no solutions, and infinite solutions. Linear equations that have one solution will be able to be transformed so that there is one variable equal to a constant. Equations that have no solutions will result in the variables adding to 0 when transformed and a false statement being left, such as $0 = 5$. Equations that have infinitely many solutions will have the variable add out when transformed, but the statement that is left will be true, such as $2 = 2$ or $0 = 0$.

Answer Key

1. $t = 200$; this means that when $t = 200$, the plans are equal and the charges would be the same.
2. Answers may vary. Be sure that the students support their answers. Answers could either reflect the current plan or Plan A, depending on how many minutes are used. Most likely, use would exceed 200 minutes, and therefore the current plan is the better choice.

3–6. Tables may vary. Be sure that Plan 1 is equal to the company's existing plan but that it differs in structure. Check that Plan 2 is either less or more economical than the existing plan always and that Plan 3 is better for some number of minutes, t. Sample answers:

	Verbal description of plan	**Algebraic expression of plan (where t is time in minutes)**
Company's existing plan	10 cents per minute plus a $50 monthly fee	$0.10t + 50$
Plan 1	25 cents per minute plus a $125 monthly fee, with a 60% discount	$0.4(0.25t + 125)$
Plan 2	10 cents per minute plus a $45 monthly fee	$0.10t + 45$
Plan 3	12 cents per minute plus a $40 monthly fee	$0.12t + 40$

7. Students' strategies may vary. Sample strategies:
Plan 1—I created a plan that uses the distributive property and makes the coefficient and the constant equal to those in the company's existing plan. This will result in infinitely many solutions when the plans are set equal to each other, thereby showing that the plans are always equal no matter how many minutes are used.
Plan 2—I created a plan that has the same rate charge per minute but a different flat fee. This will result in no solutions, where one plan is always more economical than the other and they are never equal.
Plan 3—I created a plan with a different rate charge and a different monthly fee. This will result in one solution where one plan is better than the other for a certain number of minutes. That number of minutes will be the solution to the two plans when they are set equal.

8. Linear equations that have one solution will be able to be transformed so that there is one variable equal to a constant. Equations that have no solution will result in the variables adding to 0 when transformed and a false statement being left, such as $0 = 5$. Equations that have infinitely many solutions will have the variable add out when transformed but the statement that is left will be true, such as $2 = 2$.

Differentiation

Some students may benefit through the use of calculators or hands-on equation kits during this task. Students could also use the virtual manipulative at the "NLVM: Algebra Balance Scales—Negatives" Web site listed in the Recommended Resources section of this task.

If students finish early, ask them to think about other services that charge a rate plus a flat fee and have them create plans according to the questions in Part 2. One option is to have students consider creating plans for text messaging packages or Internet services that result in comparisons with one solution, no solutions, or infinitely many solutions.

Technology Connection

Students could use a graphing calculator to graph each equation and find the intersection point(s). Students could also use a spreadsheet to test various solutions.

Choices for Students

Following the introduction, offer students the opportunity to brainstorm other services that charge a constant rate and a flat fee for use and allow them to work with a different context for solving equations that result in one solution, no solutions, or infinite solutions. Students could then choose which service they want to analyze.

Meaningful Context

The United States has several cell phone service carriers with many plans. Determining which plan has the best value can be time consuming and confusing, but millions of Americans own a cell phone and must tackle this issue. This task asks students to find when cell phone plans are equal and make meaning out of an equation that has no solutions and infinitely many solutions. The task could be expanded to have students create their own plans such that:

- two are identical in charges but different in structure;
- two will never be equal; and,
- one will be more economical after a certain number of minutes are used.

In most plans there is a charge for each minute of airtime used over the allotment of minutes, plus a monthly fee. This means that the plans are structured as $y = \text{rate}(x) + \text{monthly fee}$.

Recommended Resources

- AT&T and Verizon iPhone Plans Compared
  www.walch.com/rr/CCTTG8iPhonePlans
  This article compares cell phone plans for the iPhone from two companies and analyzes the plans.
- NLVM: Algebra Balance Scales—Negatives
  www.walch.com/rr/CCTTG8LinearEquations
  Students can create their own linear equations and solve them using the virtual manipulative at this site. The applet supports equations that have infinitely many solutions, but will tell students when they have a created an equation that has no solutions. Students can also solve equations that are given rather than creating their own.
- Purplemath.com: Solving Equations
  www.walch.com/rr/CCTTG8EquationSolutions
  This page gives worked examples of equations with one solution, no solutions, and infinite solutions. Students can use these examples to help them understand the structure of linear equations with one solution, no solutions, and infinite solutions.

8.EE.7(a) Task • Expressions and Equations
Cell Phone Plans

Part 1

A cell phone company offers a comparable airtime plan to your current plan. You are considering changing plans. Compare the two plans. Which plan would you choose and why?

	Verbal description of plan	Algebraic expression of plan (where t is time in minutes)
Your current plan	You pay 20 cents per minute plus a fee of $45 per month.	$0.2t + 45$
Plan A	You pay 25 cents per minute plus a fee of $35 per month.	$0.25t + 35$

1. Compare your current plan to Plan A by setting them equal. What is the solution? ____________________ What does the solution represent in terms of the context of this problem?

2. Which plan should you choose? ____________________ Explain your thinking and use your solution from question 1 to support your answer.

8.EE.7(a) Task • Expressions and Equations
Cell Phone Plans

Part 2

You are the owner of a cell phone company and want to offer three new plans to your existing customers. Create these plans so that, when compared to the company's existing plan, there is a plan that has one solution, a plan that has no solutions, and a plan that has infinitely many solutions. Read each question that follows and then fill out your answers in the table below to show what your proposed plans are.

	Verbal description of plan	Algebraic expression of plan (where *t* is time in minutes)
Company's existing plan		
Plan 1		
Plan 2		
Plan 3		

3. Create a model for a cell phone plan that charges a certain amount of money per minute and a flat fee per month. Add this information to the top row of the table. This is your company's existing plan.

4. Plan 1 offers a discount for the first year. Create this plan so that after the discount it is equal to the existing plan. Add this plan information to the second row of the table.

5. Plan 2 is either always better or always less economical than the existing plan. Create this plan and fill in the third row of the table.

continued

8.EE.7(a) Task • Expressions and Equations
Cell Phone Plans

6. Plan 3 has one solution where one of the plans is clearly better than the other after a certain amount of minutes as compared with the existing plan. Add this information to the last row of the table.

7. Explain your strategies for creating your plans. Detail your thinking for each plan.

8. Explain what the structures of the equations look like when they have one solution, no solutions, or infinite solutions.

Gamers for Life

Common Core State Standard

Understand congruence and similarity using physical models, transparencies, or geometry software.

8.G.2. Understand that a two-dimensional figure is congruent to another if the second can be obtained from the first by a sequence of rotations, reflections, and translations; given two congruent figures, describe a sequence that exhibits the congruence between them.

Task Overview

Background

Students often struggle with recognizing congruency in two-dimensional figures, especially those that have undergone a series of transformations. This task helps students internalize congruency of two-dimensional figures, while focusing on transformations of rotations, reflections, and translations.

The task also provides practice with:

- plotting points on a coordinate plane
- calculating horizontal and vertical distances on a coordinate plane

Implementation Suggestions

Students should work individually to complete Part 1 of the activity and should then work in pairs to complete Part 2. Students should then meet in small groups to compare and discuss their work.

Students will need rulers and may want to use colored pencils to mark new images on the coordinate planes. Consider securing access to a computer lab if you want students to play the Foldit game.

Introduction

Introduce the task by asking students if they are familiar with proteins and where proteins exist. Ask students if they are familiar with how medicines work with proteins to cure illnesses and diseases. Ask students if they are aware of any illnesses or diseases that scientists have had difficulty curing. Question students as to whether they have ever played video games. Finally, ask students how they think video games could help scientists find cures for illnesses and diseases.

Explain that our bodies contain more than a million proteins that are essential for human activity. Each protein is made up of a chain of amino acids that continues to bend and fold. Sometimes there are mistakes in the chain of amino acids that cause diseases. One disease that has baffled scientists for years is the Acquired Immune Deficiency Disorder (AIDS). In order to find a cure for AIDS, scientists must be able to understand the behavior of proteins within the human body. But it takes a great deal of time and money to find the correction needed for a cure. To help determine the folds in amino acid chains and find cures faster, scientists created the online game Foldit. Gamers are given a series of figures used to represent protein structures and are assigned transformations for each figure. With the completion of each transformation, gamers are able to create new figures, or proteins, that scientists can then use to develop drugs to fight illnesses and diseases. Now, anybody using a computer can help scientists cure diseases such as AIDS, cancer, and Alzheimer's.

Students will be performing the transformations of rotations, reflections, and translations on figures representing proteins.

Monitoring/Facilitating the Task

Ask questions and prompt student thinking so that they:

- Defend their responses. Make sure students articulate how they determined their answers.
- Prompt for and encourage the use of proper mathematic terms.
- Translate the proteins more than the given number of units to ensure variety in responses.
- Circulate and identify students with different responses to the first question.
- Question students about the meaning of rotating the protein counterclockwise 90° about the origin. Be sure they are aware of counterclockwise versus clockwise and the meaning of 90°.
- Encourage students to explore the effects of rotations, reflections, and translations on a figure if they are struggling with Part 2.
- Question students about the meaning of congruency.

Debriefing the Task

- Upon completion of the task, pairs of students should share their results within small groups.
- Select students with differing answers to share their graphs with the class. Encourage students to explain why their graphs were the same or different from one another.
- Ask students if it is possible to perform a different series of transformations and have the same result.
- Ask students whether the protein figures are still congruent if the graphs appear different. Encourage students to defend their responses.
- Prompt students to describe the process of long division with decimal numbers.
- Probe for student understanding of congruency.
- Ask students if they can think of a transformation that results in figures that are not congruent.

Answer Key

1–3. Answers may vary. Check student graphs for accuracy.

4. Yes, the proteins are congruent because the lengths of the sides and the angles of the vertices were not altered.
5. Yes, even in a different order, the new protein would still be congruent to the old protein because the lengths of the sides and the angles of the vertices were not altered.
6. The protein was translated 10 units to the right and 2 units down.
7. The protein was reflected across the *x*-axis.
8. The protein was rotated 180° about the origin.
9. The protein was reflected across the *x*-axis and then translated up 2 units.
10. Yes, the transformed proteins are congruent to the original proteins because the lengths of the sides and the angles of the vertices were not altered.

Differentiation

Some students may benefit from the use of tracing paper or transparencies for conducting the transformations as well as checking for congruency.

Technology Connection

Students who finish early can play the Foldit game themselves, using the URL listed in the Recommended Resources. Students could also use the Web site, "Transformations: Hitting a Target," found in the Recommended Resources to perform various transformations on a figure to understand the effects.

Choices for Students

Following the introduction, offer students the opportunity to create their own set of instructions for transforming a protein. Students could also create a protein and perform and record a series of transformations. Students could then share their original and transformed protein with other students and have them determine the series.

Meaningful Context

This task uses a real-world example of the online game Foldit, which is used by people with little or no background in biology to help scientists solve challenging problems. The online game makes use of the transformations of rotation, reflection, and translation to allow gamers to manipulate the three-dimensional protein. Although the actual game is three-dimensional and the task is two-dimensional, the mathematics is fundamentally the same. Most students have played video games and are relatively unaware of the mathematics involved in making them work.

Recommended Resources

- 2-D Shapes—Transformations Game
  www.walch.com/rr/CCTTG8Transformations
  Students can use this interactive site as an introduction to or review of reflections, rotations, and translations.
- CNN Tech: "Why video games are key to modern science"
  www.walch.com/rr/CCTTG8VideoGames
  This article provides background information about the University of Washington program, Foldit, which enabled gamers to produce a model of an AIDS enzyme that scientists had been trying to create for 10 years.
- Foldit: Solve Puzzles for Science
  www.walch.com/rr/CCTTG8Foldit
  This site for the original Foldit game provides FAQs and background information about the project, along with free downloads of the game (playable online or offline).
- Transformations: Hitting a Target
  www.walch.com/rr/CCTTG8TestTransformations
  Students can use this site to test the results of transforming a figure on a coordinate plane. Students can perform one transformation or a series of transformations.
- Transformations: Reflections, Translations, and Rotations
  www.walch.com/rr/CCTTG8Reflections
  The teacher or students can use this site to review reflections, translations, and rotations. Included are videos for each transformation as well as a virtual manipulative.

8.G.2 Task • Geometry
Gamers for Life

Part 1

Each protein in your body is made up of chains of amino acids that bend and fold. Mistakes in these bends and folds, called transformations, cause diseases such as cancer. Scientists can change the way a protein folds and discover new cures. However, if changing the folds also causes the protein to change *shape*, illness can result.

Below is an illustration of a protein. Scientists think they've discovered the right transformations for this protein to find a new cure, but they're not sure whether the transformations will cause the protein to change its shape. Perform the transformations below to help the scientists determine whether they've found the solution.

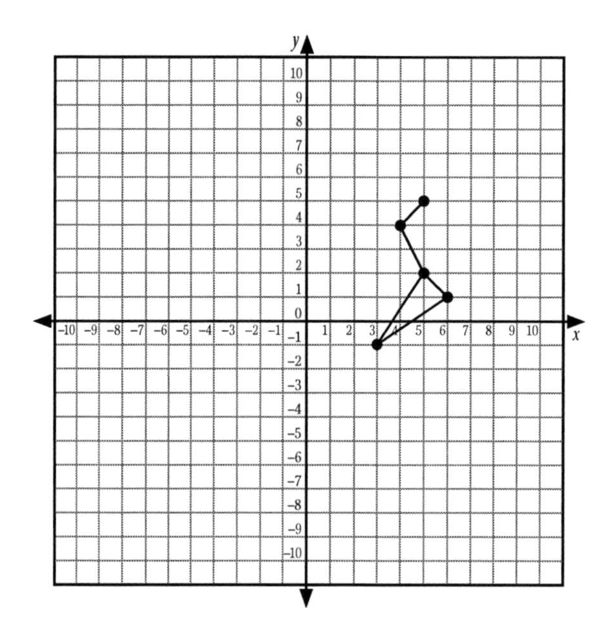

1. First, translate the protein at least 3 units horizontally and 2 units vertically.

2. Rotate the protein counterclockwise 90° about the origin.

3. Next, reflect the protein over the y-axis.

continued

4. Is the transformed protein congruent to the original protein? Explain your answer.

5. If you performed the transformation in a different order, would the new protein be congruent to the original protein? Explain your answer.

continued

8.G.2 Task • Geometry
Gamers for Life

Part 2

A computer game called Foldit is actually a tool scientists created so that people who aren't scientists can help discover new protein shapes and possible cures. A video gamer made a surprising breakthrough and discovered how some diseased proteins duplicate and cause a certain illness to spread. Scientists need to retrace the steps the gamer took to solve the problem. What transformations did the gamer use? For each problem, explain the transformations shown in each image.

6.

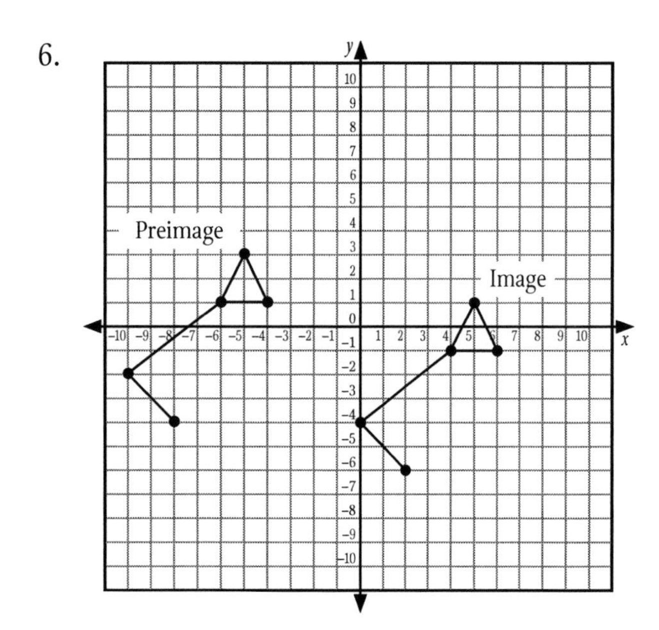

7.

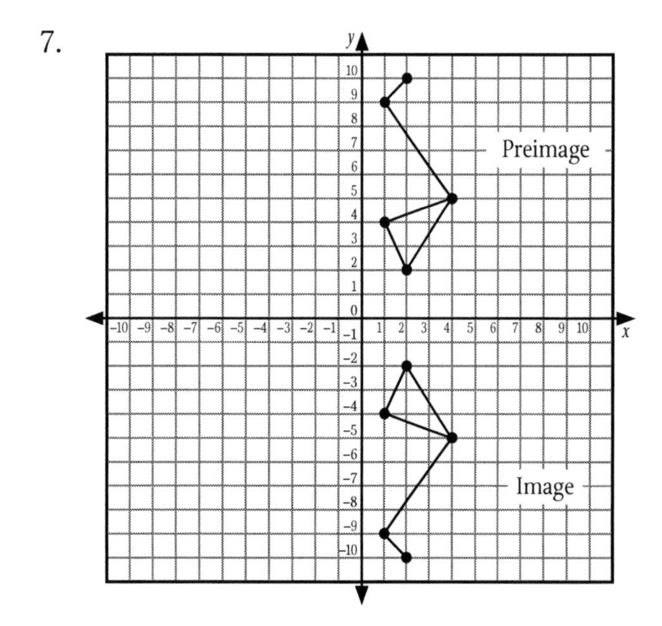

8.G.2 Task • Geometry
Gamers for Life

8.

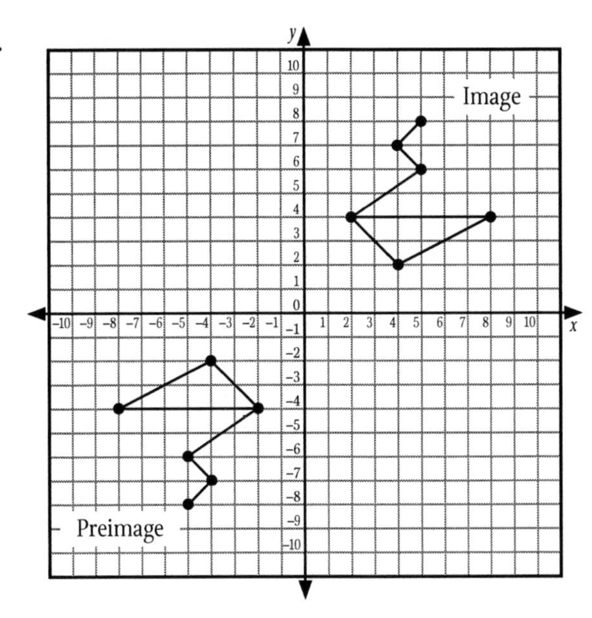

9.

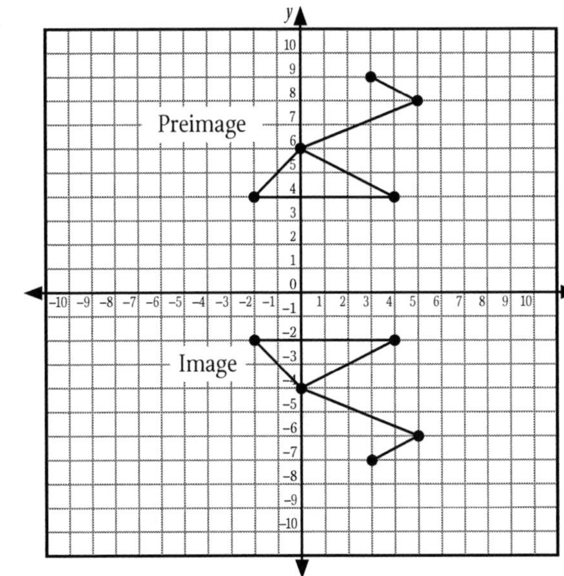

10. Are the transformed proteins congruent to the original protein? Explain how you know.

8.G.7 Task • Geometry

Get the Robot on the Stage

Instruction

Common Core State Standard

Understand and apply the Pythagorean Theorem.

8.G.7. Apply the Pythagorean Theorem to determine unknown side lengths in right triangles in real-world and mathematical problems in two and three dimensions.

Task Overview

Background

In this task, students will design a model ramp for a robot to travel from ground level onto a stage. Students will be given constraints based on the speed of the robot and the ramp slope. They will be expected to design an appropriate ramp and justify the effectiveness of the ramp. As an optional activity, they may build a scale model of the ramp they designed.

Students should have experience with using the Pythagorean theorem to calculate missing sides of a triangle, calculating slope, converting between different measures of length, and translating percent into a decimal value.

The task also provides practice with:

- calculating slope
- creating a scale model (in optional activity)

Implementation Suggestions

Students can work in groups or individually to complete this task. Calculators should be available for use during the task.

Constructing 3-D ramp models is optional. If desired, have cardboard or stiff paper, rulers, scissors, and tape available for students to use to construct the model ramp. If students construct models, it may be helpful to set up a model of the stage using the scale given in the task. Students can test their model against the stage.

Introduction

Ask students to describe where and for what purpose they have seen ramps used in their community. Set the context of the problem. Ask students if they have seen a robot and, if so, have them describe the robot's purpose. A possible answer is the Roomba®, a robot that navigates around the items in a room to vacuum the floor without human direction.

Monitoring/Facilitating the Task

Ask questions and prompt student thinking so that they:

- Focus on the constraints of the problem when designing their ramp.
- Calculate how long the ramp distance (hypotenuse) should be to get the robot on stage in 30 seconds, and then in 60 seconds.
- Use the Pythagorean theorem to calculate possible horizontal lengths based on the range available for the hypotenuse.
- Use the formula for slope to calculate the minimum horizontal distance of the ramp.
- Calculate the slope of the ramp using the possible horizontal lengths found using the Pythagorean theorem.
- Understand that the ramp must form a right triangle and therefore their side lengths must satisfy the Pythagorean theorem.
- Realize there is more than one appropriate answer to this problem.
- Select appropriate horizontal and hypotenuse lengths to form a right triangle.
- Create 3-D scale models of their ramps, if desired.

Debriefing the Task

- Discuss with the class how they determined the possible lengths for the ramp and horizontal distance. Ask them to describe how they settled on the values they used, given the range of possible values for each length. Make sure that the idea of a "good" square root is raised—using the Pythagorean theorem and knowing that the horizontal length needs to be the square of the ramp length minus 10,000 (the square of the stage height, 1 meter or 100 cm).
- Ask students to consider manipulations of the ramp constraints. Changing the slope requirements, timing requirements, or stage height would change the ramp. Ask students to describe how the ramp would change if the timeline, slope, or stage height increased or decreased.
- If Part 3 is completed, students should present their 3-D models along with justification of their models to the class. As a group, the class can agree or disagree that the model meets the constraints. Models can be left in the front of the room for comparison purposes.

Answer Key

1. Designs will vary, but the ramp length must fall between 150 cm and 300 cm. The horizontal distance must be greater than or equal to 111.8 cm. The stage height, horizontal distance, and ramp length must satisfy the Pythagorean theorem.
2. Sketches will vary; check that students appropriately labeled the lengths.
3. Answers will vary, but must fall between 15 and 30 seconds.
4. Answers will vary, but must be less than or equal to 40%.
5. Yes; answers will vary, but students must show that their values satisfy the Pythagorean theorem.
6. Answers will vary, but students must cite an appropriate time and slope.
7. Answers will vary, but must include all dimensions drawn to scale.
8. Optional 3-D models will vary; check that students adhered to the scale.

Differentiation

Some students may benefit from the assignment of different constraints. Constraints may be either less complex (the robots have upwards of 1 minute to reach the stage) or more complex (the robot moves 1 mile in 264 minutes).

Technology Connection

There are many programs available (such as Microsoft Robotics Developer Studio) to help students design and program their own robot.

Choices for Students

Students may choose to select their own ramp location or redesign an existing ramp in their community.

If allowed to pursue their "alternate" assignment, students should understand the purpose of the ramp and present the constraints on their ramp design as well as 2-D and 3-D models of their ramp.

Meaningful Context

Ramps are commonly associated with providing handicapped access to locations above or below ground. Students can find out the appropriate slope for a handicapped ramp, research the issue of inappropriately designed ramps, and explore their school and community for ramps that do not meet the standards. They may also design an appropriate ramp for the space and share their design with the owners of the ramp.

Recommended Resources

- Illuminations: Corner to Corner
  www.walch.com/rr/CCTTG8Diagonals
  This site features activities to explore diagonals and the Pythagorean theorem.
- Illuminations: Proof Without Words—Pythagorean Theorem
  www.walch.com/rr/CCTTG8PythagoreanTheorem
  This site provides a dynamic, geometric proof of the Pythagorean theorem.
- Lego Robotics
  www.walch.com/rr/CCTTG8LegoRobots
  Lego provides robot construction kits and companion software for programming robot behavior.
- Microsoft Robotics Developer Studio
  www.walch.com/rr/CCTTG8RobotDeveloper
  Download free robotics software.
- Society of Robots
  www.walch.com/rr/CCTTG8SocietyOfRobots
  The Society of Robots Web site contains a plethora of robot information and resources, including instructions for building your own robot.
- Wheelchair Ramp Safety & Standards
  www.walch.com/rr/CCTTG8WheelchairRamps
  This Web site describes the slope and length requirements for handicapped access ramps.

8.G.7 Task • Geometry
Get the Robot on the Stage

Introduction

The Robotics Club at your school has designed a new robot. They will be holding a presentation to introduce their robot in the school auditorium. They want the robot to walk onto the stage during the presentation but there is a problem—their robot cannot walk up stairs.

Your task is to design a ramp to allow the robot to walk from the floor to the stage. The design constraints are outlined as follows:

- The stage is 1 meter above the ground.
- The robot walks at a rate of 10 centimeters per second.
- The robot will have between 15 seconds and 30 seconds to reach the stage.
- The robot cannot climb a slope greater than 40% (slope is the height of the stage divided by the distance to the bottom of the ramp).
- The ramp will be 1 meter wide when built.

Part 1

1. Use the information provided to design a ramp that meets the requirements. Show your work.

8.G.7 Task • Geometry
Get the Robot on the Stage

2. Sketch a plan for your ramp to the right of the stage diagram below. Make sure to label the length of the horizontal distance from the stage to the bottom of the ramp, and the length of the ramp. The height of the stage has been included in the diagram. This model should **not** be drawn to scale.

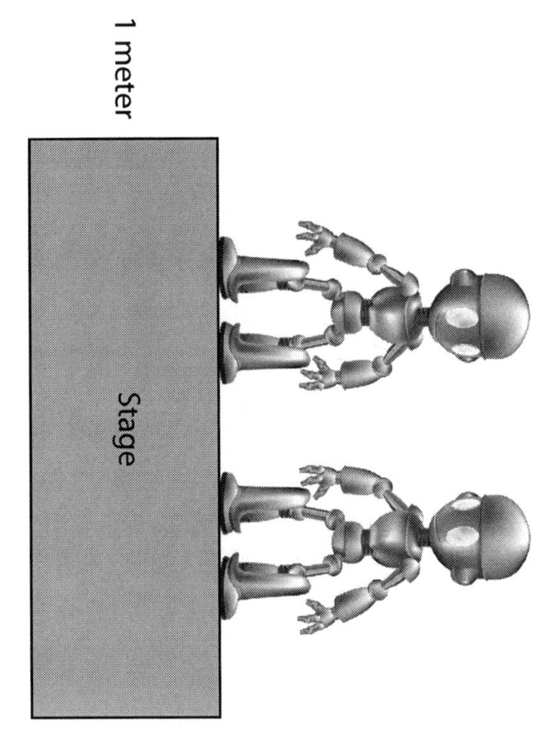

8.G.7 Task • Geometry
Get the Robot on the Stage

Part 2

3. How long will it take the robot to reach the stage using your ramp? Include your calculations.

4. What is the slope of your ramp? Show your work.

5. Is your ramp constructed as a right triangle? Justify your response.

6. Does your ramp meet the constraints from the task? Justify your response.

7. Using a ruler and a scale of 1 cm = 20 centimeters, draw a 2-D scale model of your ramp below. Label the measurements.

Part 3 *(Optional)*

8. Use the 2-D model you created to build a 3-D model of the ramp. Use the same scale you used in your 2-D model from problem 7.

8.G.9 Task • Geometry

Out of This World

Instruction

Common Core State Standard

Solve real-world and mathematical problems involving volume of cylinders, cones, and spheres.

8.G.9. Know the formulas for the volumes of cones, cylinders, and spheres and use them to solve real-world and mathematical problems.

Task Overview

Background

Students often struggle when working with volume of three-dimensional geometric solids even if they can easily calculate the area of two-dimensional figures. This task helps students internalize volume formulas as they relate to the volume of space shuttle external fuel tanks.

The task also provides practice with:

- performing operations with decimals
- identifying composite solids

Implementation Suggestions

Students may work individually or in pairs to complete one or both parts of the task. Alternatively, students may meet in groups to share their results and reflect after individually completing the task, before a class-wide discussion.

Students may want to sketch and label the geometric solids found in the shuttle on a separate sheet of paper before making any calculations.

Introduction

Introduce the task by asking students if they have ever watched a space shuttle take off and, if so, what they remember about the launch. It may be beneficial to show a short clip or photos of a space shuttle launch—see the NASA Web site links provided in the Recommended Resources for appropriate images.

Draw attention to the rockets and the large orange fuel tank. All of the fuel needed for the shuttle to travel into space is contained in the one fuel tank. Tell students the amount of fuel needed depends on the overall weight of the shuttle. Ask students how engineers might use geometry to determine the amount of fuel the tank can hold as well as the size of the tank needed for each mission.

Monitoring/Facilitating the Task

Ask questions and prompt student thinking so that they:

- Recognize that the 2-D drawing represents 3-D figures—a hemisphere, a cone, and a cylinder.
- Recognize the correct dimensions for calculating the volume of each object as well as the total volume of the external tank.
- Correctly identify the formula needed for each calculation.
- Recognize the mathematical operations they are using. Make sure that students articulate where and when they are using addition, subtraction, multiplication, and division during each calculation.
- Recognize that the end cap of the external tank is a hemisphere, or half of a sphere, and in order to calculate the volume of the end cap they will need to either divide the volume by 2 or multiply it by $\frac{1}{2}$.
- Correctly follow the order of operations.
- Understand that when they are determining the amount of fuel the external tank could hold they are calculating volume.
- Recognize that answers to volume questions should be in cubic units.
- Can defend their responses. Make sure that students articulate how they used their calculations to answer the questions.

Debriefing the Task

- When calculating volume, students should have recognized which formula and dimensions were to be used in each calculation.
- Select a few students to explain their process for determining the volume of the nose cap. Encourage students to describe how they determined which dimensions to use as well as which formula to use.
- Select students who used different methods to explain their process for determining the volume of the end cap. It may be necessary to discuss with students why multiplying the formula by $\frac{1}{2}$ is equivalent to dividing the formula by 2.
- Choose several students to volunteer and demonstrate their process for finding the total volume. Students may be concerned about answers varying by a few decimal places. It is important to discuss rounding as well as to compare answers arrived at by those who multiplied by the pi symbol on a calculator and those who multiplied by 3.14.

- Make sure that students report their calculated volumes with the appropriate units of cubic meters (m^3).
- Encourage students to use correct terms during their explanations.
- Have students share any difficulties they had completing each problem. Ask them to offer advice for overcoming those difficulties.
- Prompt students to defend their responses. Make sure that students articulate how they used their calculations to answer the questions.

Answer Key

1. a hemisphere (half a sphere), a cylinder, and a cone
2. $395.136\pi \approx 1241.36$ m^3 (1240.73 m^3 if students used 3.14); check to ensure students appropriately showed their work.
3. $395.136\pi \approx 1241.36$ m^3 (1240.73 m^3 if students used 3.14); check to ensure students appropriately showed their work.
4. $3{,}344.544\pi \approx 10{,}507.20$ m^3 (10,501.87 m^3 if students used 3.14)
5. Find the sum of the volume of the hemisphere (half sphere), the volume of the cylinder, and the volume of the cone.
6. The volume of the end cap is equal to the volume of the nose cap. The volume of the nose cap is $\frac{1}{3}(\pi)r^2h$. The height of the nose cap is twice the radius, or $\frac{1}{3}(\pi)r^2(2r)$, which equals $\frac{1}{3}(\pi)2r^3$ or $\frac{2}{3}(\pi)r^3$. The volume of the end cap is one-half of $\frac{4}{3}(\pi)r^3$ or $\frac{2}{3}(\pi)r^3$.
7. If a sphere and a cone have the same radius, then the volume of the sphere will be twice the volume of the cone.
8. If a sphere has a radius, r, then its "height" is also r and the volume is $\frac{4}{3}(\pi)r^3$. The volume of a cone with height r is $\frac{1}{3}(\pi)r^2h$ or $\frac{1}{3}(\pi)r^3$, and the volume of a cylinder with height r is πr^2h or πr^3. Therefore, the volume of a cylinder is 3 times the volume of a cone, and the volume of a sphere is 4 times the volume of a cone.

Differentiation

Some students may benefit from the use of calculators during this task. Some may need a list of geometric solids as well as visuals for each solid available when beginning the task. You may wish to provide the formulas for calculating the volume to students who need them.

Students who finish early could research the history of the space program as well as the design changes over the years. Students could calculate the volume of the external tank in each design.

Students could create a spreadsheet to compare the volume of various-sized tanks by having one variable remain constant.

Technology Connection

After completing the task, students could access the "Volume" Web site listed in the Recommended Resources to experiment with changing the radius and height of a cylinder, a cone, and a sphere.

Choices for Students

Following the introduction, offer students the opportunity to research past shuttle designs as well as proposed future designs. Students could calculate the volume of the external tank in each design. Students could also design their own shuttle fuel tank and calculate the volume. This task may be expanded by having students research model rocketry and methods used for launching. Have students calculate and compare the volumes of the different rocket cylinders. Students could also design their own rocket and calculate the volume of the cylinders.

Meaningful Context

NASA's shuttles traveled to outer space by the force of two solid rocket boosters and the fuel of one external tank. The amount of fuel the shuttle carried was based on the weight of the shuttle. The shuttle program has been discontinued, but private entrepreneurs are racing to launch a "space tourism" industry. Private citizens could pay money to fly into orbit. To date, only the Russian Space Agency has taken paying customers into space. In America, Virgin Galactic is building a spaceport in New Mexico and accepting reservations for flights into space. See the Recommended Resources for links to Web sites with further information.

Recommended Resources

- Cylinders, Cones, and Spheres
  www.walch.com/rr/CCTTG8CylindersConesSpheres
  This site includes an overview of cylinders, cones, and spheres. Students can continue through the pages of the site (see the four "steps" at the top of the site page) to review formulas and examples for finding the volumes of cylinders, cones, and spheres.
- NASA Announces Design for New Deep Space Exploration System
  www.walch.com/rr/CCTTG8SpaceLaunch
  This Web site provides information about the new Space Launch System for carrying crew and cargo to space. An artist's rendition of a takeoff is also included.
- NASA: Space Shuttle Launch and Landing
  www.walch.com/rr/CCTTG8SpaceShuttle
  This site contains information and select, high-quality images of NASA's space shuttle launches and landings.
- Virgin Galactic
  www.walch.com/rr/CCTTG8VirginGalactic
  This private company, an offshoot of Virgin Atlantic Airways, is building a spaceport in New Mexico and selling reservations for a planned flight into space. Students can explore the interactive timeline and read about the types of rockets used by the company's spaceships.
- Volume
  www.walch.com/rr/CCTTG8VolumeApplets
  This site contains links to interactive applets that allow students to change the heights and radii of cones, cylinders, and spheres.

NAME:

8.G.9 Task • Geometry
Out of This World

Introduction

NASA's space shuttles traveled to outer space by the force of two solid rocket boosters and the fuel of just one external tank. The amount of fuel that the shuttle carried was based on the weight of the shuttle. What is the maximum amount of fuel that the external tank of the space shuttle could hold? A representation of the space shuttle's external fuel tank is below.

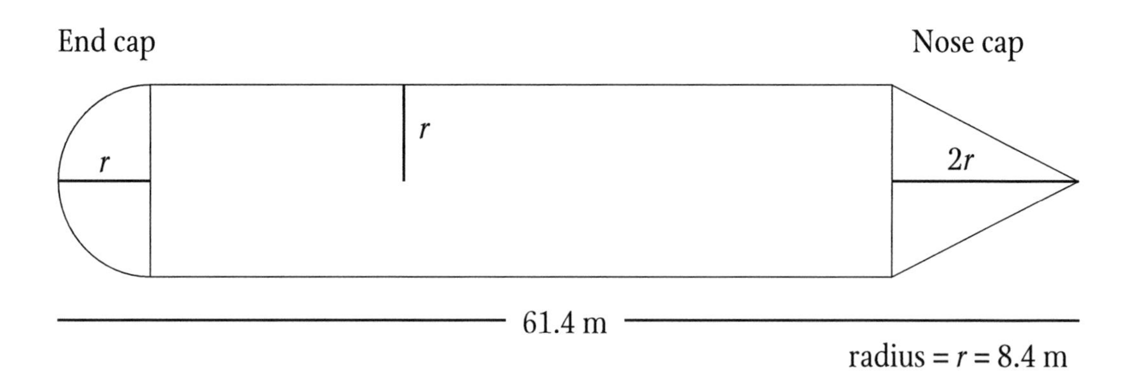

Task Questions

1. Look at the diagram above. What 3-D geometric solids make up the structure of the tank?

2. The height of the nose cap is equal to double the radius of the external tank. If the radius of the external tank is 8.4 meters, what is the volume of the nose cap? Show your work.

3. The radius of the external tank is equal to the radius of the end cap. What is the volume of the end cap? Show your work.

continued

8.G.9 Task • Geometry
Out of This World

4. What is the total volume of the space shuttle's external tank? Show your work.

5. Explain your process for finding the total volume of the external tank.

6. Review the answer you got when you calculated the volume of the end cap. Compare this answer to your volume calculation for the nose cap. What do you notice? Explain.

7. What does your answer to question 6 tell you about the relationship between the volume of a sphere and the volume of a cone if their radii are equal?

8. Now consider the volume you calculated for the cylinder. What is the relationship among the volumes of these three geometric solids—a sphere, a cylinder, and a cone—if each has the same height and radius?

Effects of Urbanization and Logging on Watersheds

Common Core State Standard

Investigate patterns of association in bivariate data.

8.SP.1. Construct and interpret scatter plots for bivariate measurement data to investigate patterns of association between two quantities. Describe patterns such as clustering, outliers, positive or negative association, linear association, and nonlinear association.

Task Overview

Background

Students often struggle when working with bivariate data despite their ability to plot accurately in four quadrants. This task helps students internalize the process of creating and interpreting scatter plots for patterns and associations while drawing their attention to the effects of development on the environment and watersheds.

The task also provides practice with:

- converting bivariate data into coordinates
- reasoning and justifying conclusions

Implementation Suggestions

For Part 1, students can work individually or in pairs. For Parts 2 and 3, split the class into two groups so that half the students work on Part 2 and half the students work on Part 3. Bring the whole class back together to debrief Parts 2 and 3 together.

Alternatively, for Parts 1 and 2, students may work individually, in pairs, or in small groups to complete one or both parts of the task. Part 3 is optional if time allows or could be used for students who finish early.

Introduction

Introduce the task by asking students about patterns, trends, and associations. Ask students to give some examples of bivariate (two-variable) data that they know exhibits patterns, trends, or associations. Students might suggest that there is an association between height and weight, education and earning potential, age and sleep, or any other various bivariate data groups. If students are having trouble coming up with examples, you might want to suggest the aforementioned associations. Ask students about their understanding of positive and negative associations and how these associations would appear on a graph. Ask students to talk about the difference in appearance between linear and nonlinear graphs. Elicit prior knowledge of water pollution and the effects of development on watersheds.

One way to measure the disturbance to the environment from urbanization and logging is to analyze the quantity and quality of macroinvertebrates in the streams near developed and undeveloped land areas.

Macroinvertebrates are animals that do not have a backbone and can be seen without the aid of a microscope. The health of such organisms is measured by their ability to thrive in their natural environment. This measurement score is called the Benthic Index of Biotic Integrity (B-IBI). For the purpose of simplifying the task to a level the students will understand, this score name has been changed to "species health score."

The "development score" used in this task refers to the level of development by humans of a given environment or area. A lower score indicates a low level of development; high scores indicate highly developed areas. The different land areas listed in Part 1 are rural, suburban, and urban areas around Puget Sound in Washington state.

Logging also disturbs watersheds. After a storm, water flows heavily into watersheds. The flow of water is affected by human factors such as development. Without trees and other plant life to soak up rainwater from storms, the water runs directly into the streams, carrying pollutants and debris. Officials for the H.J. Andrews Experimental Forest in Blue River, Oregon, studied the impact of water discharge from storm events on logged and un-logged areas to determine the effect of clear-cutting and road construction on watersheds. Those data are used in Part 2. The water flow into the streams is measured in cubic meters per second over the land mass area.

Monitoring/Facilitating the Task

Ask questions and prompt student thinking so that they:

- Recognize which data set goes on the *x*-axis and which data set goes on the *y*-axis. The best choice is for "Development score" to go on the *x*-axis and the total "Species health score" to go on the *y*-axis because, in this case, "Development score" is the explanatory (independent) variable and "Species health score" is the response (dependent) variable. It is possible that the axes could be reversed.
- Plot points accurately and choose appropriate scales. An appropriate scale for the *x*-axis is from 0 to 70 in increments of 10. The *y*-axis scale could go from 0 to 50 in increments of 5. Scales may vary slightly. Encourage students to use the entire grid provided.
- Explain their reasoning in determining patterns, associations, clustering, and outliers.

- Identify the difference between a linear and nonlinear pattern.
- First look at the data mathematically and then supply the contextual meaning.
- Support their conclusions with arguments based on the data.

Debriefing the Task

Part 1

- Students will be creating and interpreting scatter plots.
- For question 1, students could have chosen either data set to place on the *x*-axis. Ask for student examples that set up the graph in each way. Ask students to explain their choice of which data set to place on the *x*-axis. If more than one student or group of students chose one variable over the other to place on the *x*-axis, encourage discussion about the reasons for choosing the *x*-axis variable. The better choice is to have "Development score" on the *x*-axis because this is what is being explained, while "Species health score" is the response to development in the area ("Development score").
- For question 2, prompt students to report out differences in describing the patterns they saw. Some students might see the data differently and can help others to look at the data differently than they originally recorded.
- In question 3, some students might see clustering while others may not. Either answer is correct, but the defense is more important than the answer. Students might suggest that there is clustering because they see the data being grouped together toward the upper left-hand corner. Other students might see that the data is spread out over the coordinate plane.
- Question 4 has a clear answer of linear. Encourage those who answered "linear" to explain their thinking.
- For question 5, encourage student discussion about the presence or absence of outliers. Some students might think that the data point (60, 9) is an outlier. While this point is slightly separated from the rest of the data, it still follows the negative association and the linear trend, so it appears as though it is not an outlier. Prompt students to think about how the pattern would change if the data point were removed; help them to see that the pattern wouldn't change. Explain that an outlier often changes the pattern or deviates from the pattern, and this data point does neither.
- Question 6 gives students the opportunity to synthesize the meaning of the first five questions in terms of the context of the problem. This question asks students to make meaning of the patterns and associations.

Part 2

- In this part, students will be interpreting scatter plots.
- Question 7 gives students the opportunity to contrast this pattern with the pattern from question 1 in Part 1. Encourage students to compare and contrast the two patterns. Both patterns are linear, but the pattern from Part 1 has a negative association, whereas the pattern in question 7 has a positive association.
- Question 8 shows some clustering at the bottom left of the graph. Encourage students to report out how they made meaning of this clustering. Prompt students to give various explanations. Some students might see that the difference between water flow in smaller areas is less drastic than in larger ones. Others might conclude that there are more areas that are smaller in surface area for comparison.
- In question 9, students will most likely report that the model is linear. If students think that the model is nonlinear, ask them to explain their thinking. The data do appear to curve slightly toward the upper right of the graph, but a linear model would still work. Encourage students to defend their responses.
- The data in question 10 could be interpreted as having or not having any outliers. If students suggested a linear model, then they might think that point (1.65, 1.8) is an outlier because the data appears to curve a bit with this point; if the point were removed, the data would better fit a straight line. However, even with that point the data follows a linear trend with a positive association. Encourage discussion among students with different answers.
- Question 11 gives students the opportunity to synthesize the meaning of the previous five questions in terms of the context of each problem. This question asks students to make meaning of the patterns and associations.
- For question 12, students are asked to interpret the meaning of the data analysis by drawing a conclusion and taking a position of whether or not to develop the land. This question synthesizes information from both Parts 1 and 2. Encourage students to make this synthesis and discuss their conclusions.
- Assess understanding by expanding the parameters used in the task by using different data sets. Other data sets can be found on the Quantitative Environmental Learning Project Web site, listed under Recommended Resources.

Part 3

- In this part, students will be interpreting scatter plots and synthesizing conclusions from Part 1 and also from Part 3 itself.
- Questions 13 and 14 clearly show a nonlinear trend in the data. Some students might recognize this as exponential decay.
- Question 15 gives students the opportunity to synthesize the data and make meaning of the data in terms of the context of the problem. This question also asks students to draw upon their conclusions from Part 1. Encourage discussion of the students' conclusions and supporting arguments.

Answer Key

1.

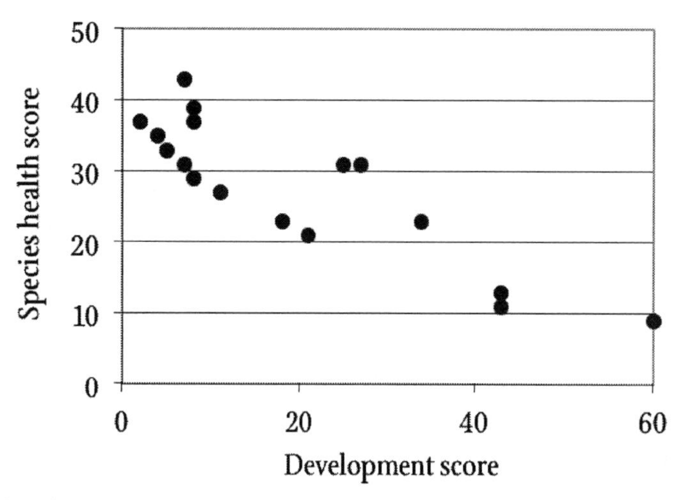

2. The data appear to decrease as the developed area increases. The data go from the top left toward the bottom right, indicating a negative association.
3. The data appear to cluster a little bit at the start of the graph in the upper left corner and then spread out toward the bottom right, but because the clustering isn't truly clear, the graph could also be interpreted as having no clustering. Be sure that students defend their response.
4. A linear model appears to be a better fit for the data than a nonlinear model since the data roughly follows a straight line going from the top left of the graph to the bottom right (negative association or negative slope).
5. There do not appear to be any outliers since all the data points roughly follow a linear model with a negative association.

6. As the development increases, the health of the water species decreases. This means that development is harming the watershed.
7. The data appear to go from the bottom left toward the top right with some clustering in the bottom left, suggesting a positive association.
8. The data appear to cluster near the bottom left corner of the graph. This means that, for smaller surface areas, the logging has less effect on the watershed and the water flow is roughly the same between logged and un-logged areas. However, since the data spreads out for larger surface areas of logged areas, there appears to be greater water flow into the watershed for those logged areas.
9. A linear model appears to be a better fit for the data than a nonlinear model since the data roughly follow a straight line going from the bottom left of the graph to the top right (positive association or positive slope).
10. There do not appear to be any outliers since all the data points roughly follow a pattern and none stick out from that positive linear association pattern. However, some students might report that the last data point in the upper right corner is an outlier since it appears to be set apart from the rest of the data. Be sure students justify their answers.
11. The data suggests, since the pattern is positive and linear, that both logged and un-logged areas experience greater water flow with more volume of water (rain). However, analyzing the scatter plot more carefully reveals that in most cases for the same amount of rain volume, there is more flow in logged areas than un-logged areas. This means that there is more water flowing into watersheds near logged areas than un-logged areas. This water could potentially be carrying pollutants.
12. Answers will vary. Be sure that students defend their answers and support their position using correct terminology. Sample answer: The town should not develop the land because this would increase the amount of storm water draining into the stream. Based on the conclusions from Part 1, this means that more pollutants would be introduced and the health of the water species would decline.
13. The data falls to the right.
14. A nonlinear model would better fit the data because the data does not follow the pattern of a straight line; it's curved.
15. The more rain there is, the more water flows into the watershed. This data is based on an un-logged area, so if the area were to be logged the water flow would increase and introduce more pollutants into the watershed. Based on the data from Part 1, increased pollutants and greater water runoff into the watershed decrease the health of the water. If the area were to be logged and the water discharge recorded, the graph would most likely shift up.

Differentiation

Some students may benefit from the use of graphing calculators and/or a spreadsheet for Part 1.

If you break up the students into two groups to complete Parts 2 and 3, struggling students might find Part 3 easier to complete because the association is more clear in Part 3 than in Part 2. If all students complete Parts 1 and 2, Part 3 may be used as an extension or given to students who finish early.

If students finish early, have them develop a PowerPoint to present to the "town council" regarding the development of a land parcel. Have students use the data and their conclusions from the task to support their position to either develop or not develop the land.

Technology Connection

Students could use a graphing calculator or a spreadsheet to create their scatter plots.

Choices for Students

Following the introduction, offer students the opportunity to analyze their own data. Several data sets can be found at the Quantitative Environmental Learning Project Web site listed in the Recommended Resources.

Meaningful Context

Urbanization is disturbing the environment. This can be seen in data collected from streams, rivers, and lakes. This data is often plotted using scatter plots for the bivariate data and analyzed for clustering, outliers, positive or negative association, linear association, and nonlinear association.

Recommended Resources

- Purplemath.com: Scatterplots and Regressions
  www.walch.com/rr/CCTTG8ScatterPlots
  This site provides a lesson on creating scatter plots and interpreting them. Examples of positive, negative, linear, nonlinear, and no associations are provided, as well as instruction for creating a line of best fit.
- Quantitative Environmental Learning Project
  www.walch.com/rr/CCTTG8EnvironmentData
  This site offers data on several environmental issues as well as data summaries and scatter plots. This resource is useful for students who wish to make alternate selections of data for studying patterns and association in scatter plots.
- U.S. Environmental Protection Agency: Biological Indicators of Watershed Health—Benthic Macroinvertebrates in Our Waters
  www.walch.com/rr/CCTTG8WatershedHealth
  This site provides pictures and links to descriptions of common macroinvertebrates found in water.

NAME:

8.SP.1 Task • Statistics and Probability
Effects of Urbanization and Logging on Watersheds

Part 1

Does building on undeveloped land affect the water supply? The table below records the amount of development around several watersheds. It also shows how healthy the species are in each watershed. You will create a scatter plot of the data. Then, you will see if the scatter plot shows any damage to the watersheds.

- "Development score" shows the amount of buildings, sidewalks, etc., in the area. Areas with more development have higher scores.
- "Species health score" represents the health of organisms found in watersheds. Healthier organisms have higher scores.

Development of Watersheds in Puget Sound

Name of watershed	Development score	Species health score
Thornton Creek #3	60	9
Kelsey Creek	43	11
Juanita Creek #3	43	13
Schneider Creek	34	23
North Creek	27	31
Swamp Creek	25	31
Coal Creek	21	21
Percival Creek #2	18	23
Percival Creek #1	11	27
Covington Creek	8	37
Little Anderson Creek	8	39
Big Beef Creek	8	29
Seabeck Creek	7	31
Rock Creek	7	43
Stavis Creek	5	33
Carey Creek	4	35
Big Anderson Creek	2	37

Data adapted from: www.walch.com/CCTTG8PugetWatersheds

NAME:

8.SP.1 Task • Statistics and Probability
Effects of Urbanization and Logging on Watersheds

1. Create a scatter plot of the data. Label the axes and add a title for your graph on the lines provided.

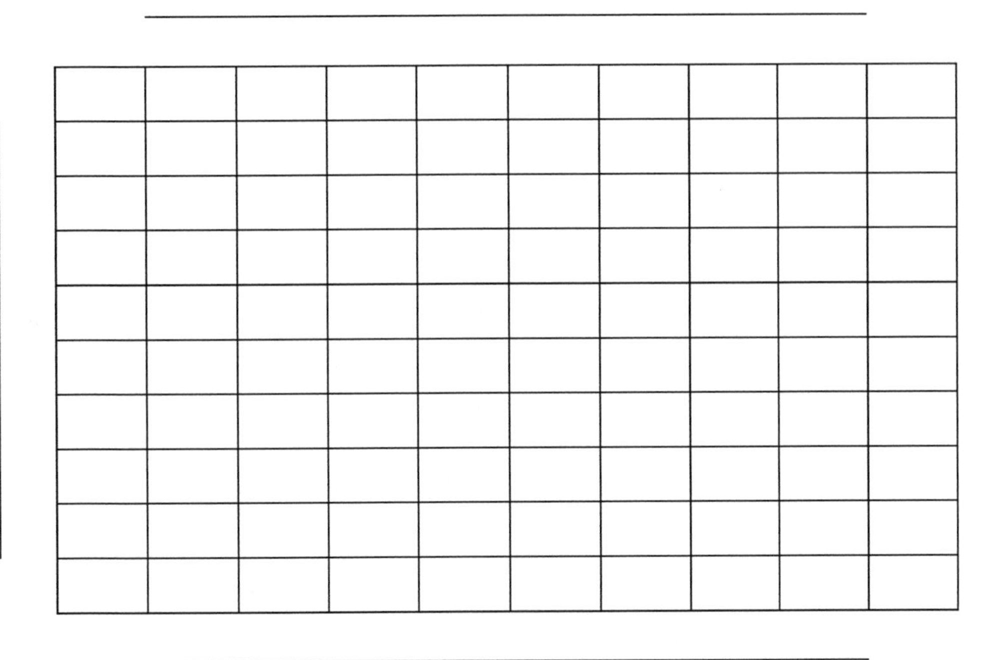

2. Do you notice any patterns in the data? Explain.

3. Does the data look like it has any clusters? Explain.

8.SP.1 Task • Statistics and Probability
Effects of Urbanization and Logging on Watersheds

4. Do you think the data follows a linear or nonlinear model? ____________________
   Explain your thinking.

5. Do there appear to be any outliers? If so, where are they?

6. What can you conclude from the data table and scatter plot?

continued

Part 2

Your town is considering a development project. Your job is to find out how the development will affect the watershed. The graph below compares how much storm water flows into watersheds in areas with trees and without trees. The x-axis represents water flow over land that hasn't been logged. The y-axis shows water flow over land that has been logged.

What can you conclude from this graph about how development affects water flow? Would you support the proposed development based on this data? Use your discoveries from Part 1 and this part to support your position.

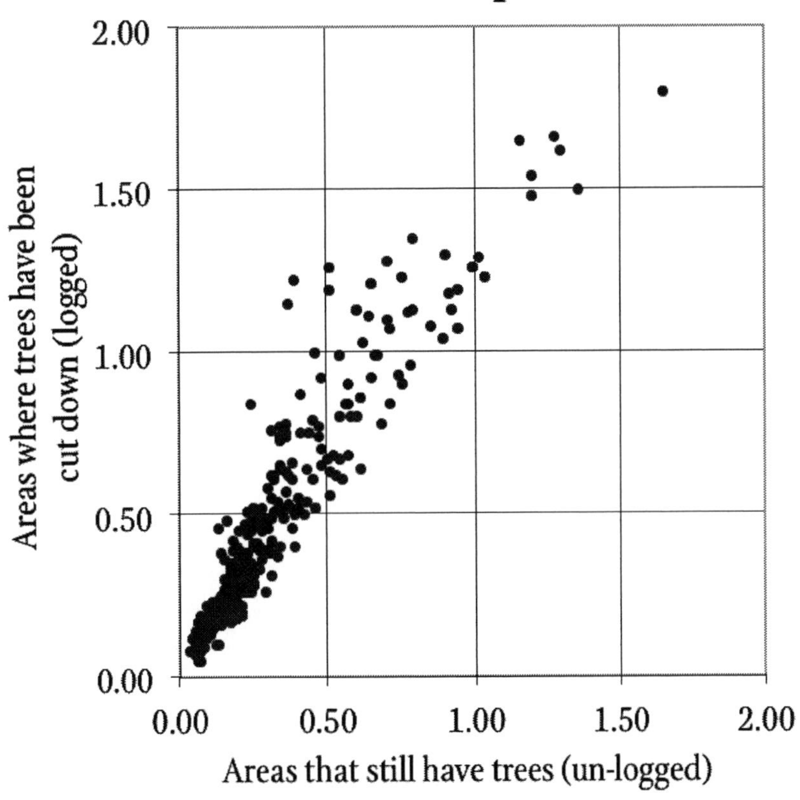

Data adapted from: www.walch.com/CCTTG8StormFlow

7. Do you notice any patterns in the data? Explain.

continued

8.SP.1 Task • Statistics and Probability
Effects of Urbanization and Logging on Watersheds

8. Does the data look like it has any clusters? Explain in terms of the context of this problem.

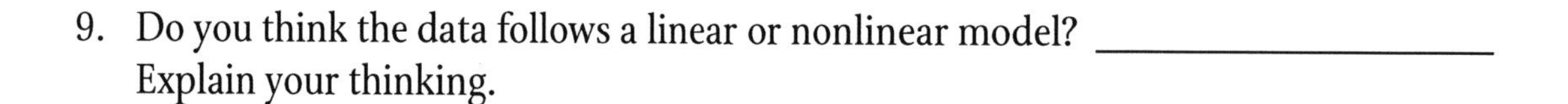

9. Do you think the data follows a linear or nonlinear model? ____________________
   Explain your thinking.

10. Do there appear to be any outliers? If so, where are they?

11. What can you conclude from the scatter plot?

12. Would you support the proposed development? Use data and conclusions from Parts 1 and 2 to support your answer.

continued

Part 3

Your town is considering a new housing development. To build it, part of the forest around the town's river would be cut down. The river is the watershed that provides the town's drinking water. You have been asked to determine how the development will affect the amount of storm water that flows into the river. In the graph below, the y-axis represents how much water ran into the river after a recent major storm. The x-axis shows the time in hours after the storm.

Based on your summary from Part 1 and the graph below, what can you conclude about the future health of the watershed if the development is approved? Use your discoveries from Part 1 and this part to defend your position.

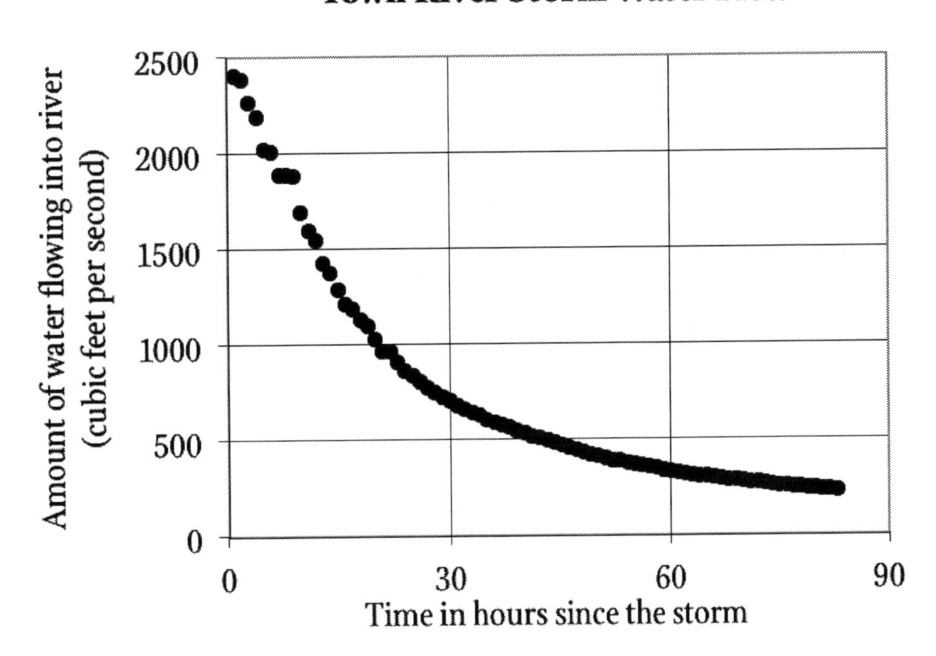

Data adapted from: www.walch.com/CCTTG8RiverStormFlow

13. Do you notice any patterns in the data? Explain.

continued

14. Do you think the data follows a linear or nonlinear model? ____________________
Explain your thinking.

15. Based on your analysis of the data from Parts 1 and 3, what can you conclude about the effect on this watershed if the land is developed? How do you think the scatter plot would change if the forest were logged?